MW00677933

Preventive Maintenance

Terry Wireman, CPMM

www.terrywireman.com
TLWireman@Mindspring.com

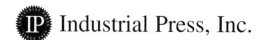

Industrial Press, Inc.

Library of Congress Cataloging-in-Publication Data
Wireman, Terry.
   Preventive maintenance / Terry Wireman.
   p. cm.(Maintenance strategy series; v.1)
   ISBN 978-0-8311-3300-9
   1. Maintenance. 2. Industrial equipment--Maintenance and repair.. I. Title.

TS192.W57.2007
658.2'02--dc22   2006053157

Industrial Press, Inc.
989 Avenue of the Americas
New York, NY 10018

First Edition, 2008

Sponsoring Editor: John Carleo
Interior Text and Cover Design: Janet Romano
Developmental Editor: Robert Weinstein

10   9   8   7   6   5   4   3   2   1

TABLE OF CONTENTS

iii

INTRODUCTION

Volume 1: Preventive Maintenance

Preventive Maintenance is the foundation of the entire maintenance philosophy. Unless the PM program is effective, all subsequent maintenance activities will:
- Cost more then they should to implement
- Take longer than they should to implement
- Have a higher probability of failure

The bottom line: Unless the PM program is successful, the rest of the maintenance functions will be sub-optimized.

Volume 1 in the *Maintenance Strategy Series* deals with the function of preventive maintenance, which is the starting point for developing a complete maintenance strategy. This function is one of the most basic for a maintenance organization; however, it is also the most ignored and underutilized function. As will be shown, this function is key to unlocking the financial benefits that can be provided by a comprehensive maintenance strategy.

Many organizations develop the "roller coaster" syndrome with preventive maintenance. They start at the bottom of the roller coaster ride when someone in the organization begins complaining about how the company's assets are performing. In the past, the company may have compensated for poor asset performance by purchasing more assets, resulting in redundant equipment. Or, the company may have decided to run additional shifts with the existing equipment, resulting in overtime charges or excess employees. In either case, the financial performance of the company would be trending downward.

Finally, someone proposes the development of a preventive maintenance program. In desperation, the organization decides that this is something they need to try. So with hard work, they hook the chain to the "roller coaster" and begin to pull it up the hill. The initial effort is difficult, but as the preventive maintenance program begins to be effective, it become easier to execute. In fact, the maintenance department begins to gain stature in the organization because the equipment is performing so well. So as the roller coaster is pulled to the top of the first hill, the scenery looks great. The company is making progress, the equipment is performing well; all is good.

What causes the first plunge? A person with a pure financial focus will ask "If our equipment is performing so well, why do we need so many maintenance people in the department?" Even though the past history can be explained, the decision is made to "downsize" the maintenance department. Now the downward part of the ride begins. Because there are fewer maintenance technicians, less preventive maintenance work takes place. Eventually, more breakdown work occurs, resulting in even fewer resources to perform preventive maintenance. At this point, the equipment performance begins to degrade. As the company's financial performance begins to suffer, someone will ask about the impact a preventive maintenance program would have on the current situation.

The decision will be made to reestablish the preventive maintenance program. Depending on how far down the hill the preventive maintenance program had gone, it may have the momentum to spring back into effectiveness quickly. If the organization had deteriorated to a completely reactive state, with some of the maintenance management having actually left the organization, it may be a time and resource consuming effort to reestablish it. In fact, just as some roller coasters have secondary (and even third chains) to reestablish the coaster's momentum, it may take considerable effort to make the preventive maintenance program successful.

Once the preventive maintenance program is successful and conditions improve, again a person with a pure financial focus will ask "If our equipment is performing so well, why do we need so many maintenance people in the department?" If the organization can not prove why the maintenance personnel are needed, the roller coaster will once again start down the hill.

Ridiculous? Yes, it is. However, there are countless organizations are stuck on this endless loop of the preventive maintenance roller coaster. Hopefully, by applying the methodology in this text, companies can stop the preventive maintenance roller coaster and move on to the next functional phase in implementing a maintenance strategy: MRO inventory and procurement. This topic will be addressed in Volume 2 of the *Maintenance Strategy Series*.

Overview

The Maintenance Strategy Series Process Flow

Good, sound, functional maintenance practices are essential for effective maintenance / asset management strategies. But what exactly are "good, sound, functional maintenance practices?" The materials contained in this overview (and the overview for each of the volumes in the Maintenance Strategy Series) explain each block of the Maintenance Strategy Series Process Flow. They are designed to highlight the steps necessary to develop a complete maintenance / asset management strategy for your plant or facility. The activities described in the Process Flow are designed to serve as a guide for strategic planning discussions. The flow diagram for the Maintenance Strategy Series Process Flow can be found at the end of this overview.

Author's Note

Many individuals may believe that this type of PM program is too expensive or time consuming to implement, especially when there are advanced predictive or reliability techniques that might be employed. Yet there is a reason for the sequencing of the Maintenance Strategy Series process flow. If attempts are made to deploy advanced techniques before the organization is mature enough to properly understand and utilize them (basically, the "I want results now" short-term focus), they will fail. The reason? Developing and implementing a sustainable maintenance / asset management strategy is more than just distributing a flow chart or dictionary of technical terms. It is an educational exercise that must change a company culture. The educational process that occurs during a structured implementation of basic maintenance processes must evolve into more sophisticated and advanced processes as the organization develops the understanding and skills necessary.

If an individual is to obtain a college degree, it may involve an investment of four or more years to achieve this goal. Likewise, if a company is to obtain an advanced standing in a maintenance / asset management strategy, it may take up to four years. It is not that someone cannot, through years of experience and education, design their maintenance / asset management process in a short time period. It will, however, take the entire organization (from senior executive to shop floor employees) this amount of time to become mature in their understanding and utilization of the process. Although there will be incremental benefits achieved along the journey to maintenance / asset management excellence, the true benefits are not realized until there is a complete organizational focus on maximizing all aspects of the investment in the assets. It is this competitive focus that separates long-term, sustainable success from a short-term "flash" of improvement.

In the beginning, it is necessary for a plant or facility to decide it is necessary to improve their maintenance / asset management strategy. The business reason for the needed improvement can be multi-faceted, but would likely include:

- Poor Return On Investment (ROI) for the total plant or facility valuation
- Poor throughput for the design of the plant
- Inability to meet production demands
- High cost of occupancy for a facility
- Excessive downtime
- Production inefficiencies

Once the decision has been made to develop / improve the maintenance/ asset management strategy, the Maintenance Strategy Series process flow diagram should be followed. It begins with Preventive Maintenance.

1. Does a PM Program Exist?
Preventive maintenance is the core of any equipment/asset maintenance process improvement strategy. All plant and facility equipment, including special back-up or redundant equipment, must be covered by a complete, cost-effective, preventive maintenance program. The preventive maintenance program must be designed to eliminate all unplanned equipment failures. The preventive maintenance program should be designed to insure proper coverage of the critical equipment of the plant or facility. The program should include a good cross section of the following:

- Inspections
- Adjustments

- Lubrication
- Proactive replacements of worn components

The goal of the program is to insure there will be no unplanned equipment downtime.

2. Is the PM Program Effective?

The effectiveness of the preventive maintenance program is determined by the level of unplanned equipment maintenance that is performed. Unplanned equipment maintenance is defined as any maintenance activity that is performed with less than one week of advanced planning. Unplanned equipment maintenance is commonly referred to as reactive maintenance. An effective preventive maintenance program will reduce the amount of unplanned work to less than 80% of the total manpower expended for all equipment maintenance activities. If more time than this is being spent on unplanned activities, then a reevaluation of the preventive maintenance program is required. It will take more resources and additional time to make progress in any of the following maintenance process areas unless the preventive maintenance program is effective enough for the equipment maintenance to meet the 80%/20% rule.

It should be the goal of not progressing any further until the preventive maintenance program is successful. In addition to requiring more resources and taking longer to develop the subsequent maintenance processes, it is very common to see companies try to compensate for a reactive organization. This means they will circumvent some of the "best practices' in the subsequent processes to make them work in a reactive environment. All this will do is reinforce negative behavior and sub-optimize the effectiveness of the subsequent processes.

3. Do MRO Processes Exist?

After the preventive maintenance program is effective, the equipment spares, stores, and purchasing systems must be analyzed. The equipment spares and stores should be organized, with all of the spares identified and tagged, stored in an identified location, with accurate on-hand and usage data. The purchasing system must allow for procurement of all necessary spare parts to meet the maintenance schedules. All data necessary to track the cost and usage of all spare parts must be complete and accurate.

4. Are the MRO Processes Effective?

The benchmark for an effective maintenance / asset management MRO process is service level. Simply defined, the service level measures what percent of the time a part is in stock when it is needed. The spare parts must be on hand at least 95%– 97% of the time for the stores and purchasing systems to support the maintenance planning and scheduling functions.

Again, unless maintenance activities are proactive (less than 20% unplanned weekly), it will be impossible for the stores and purchasing groups to be cost effective in meeting equipment maintenance spare parts demands. They will either fall below the 95%–97% service level, or they will be forced to carry excess inventory to meet the desired service level.

The MRO process must be effective for the next steps in the strategy development. If the MRO data required to support the maintenance work management process is not developed, the maintenance spare parts costs will never be accurate to an equipment level. The need for this level of data accuracy will be explained in Sections 6 and 10 of the preface.

5. Does a Work Management Process Exist?

The work management system is designed to track all equipment maintenance activities. The activities can be anything from inspections and adjustments to major overhauls. Any maintenance that is performed without being recorded in the work order system is lost. Lost or unrecorded data makes it impossible to perform any analysis of equipment problems. All activities performed on equipment must be recorded to a work order by the responsible individual. This highlights the point that maintenance, operations, and engineering will be extremely involved in utilizing work orders.

Beyond just having a work order, the process of using a work order system needs detailed. A comprehensive work management process should include details on the following:
- How to request work
- How to prioritize work
- How to plan work
- How to schedule work
- How to execute work
- How to record work details
- How to process follow up work
- How to analyze historical work details

6. Is the Work Management Process Effective?

This question should be answered by performing an evaluation of the equipment maintenance data. The evaluation may be as simple as answering the following questions:

- How complete is the data?
- How accurate is the data?
- How timely is the data?
- How usable is the data?

If the data is not complete, it will be impossible to perform any meaningful analysis of the equipment historical and current condition. If the data is not accurate, it will be impossible to correctly identify the root cause of any equipment problems. If the data is not timely, then it will be impossible to correct equipment problems before they cause equipment failures. If the data is not usable, it will be impossible to format it in a manner that allows for any meaningful analysis. Unless the work order system provides data that passes this evaluation, it is impossible to make further progress.

7. Is Planning and Scheduling Utilized?

This review examines the policies and practices for equipment maintenance planning and scheduling. Although this is a subset of the work management process, it needs a separate evaluation. The goal of planning and scheduling is to optimize any resources expended on equipment maintenance activities, while minimizing the interruption the activities have on the production schedule. A common term used in many organizations is "wrench time." This refers to the time the craft technicians have their hands on tools and are actually performing work; as opposed to being delayed or waiting to work. The average reactive organization may have a wrench time of only 20%, whereas a proactive, planned, and scheduled organization may be as high as 60% or even a little more.

The ultimate goal of planning and scheduling is to insure that all equipment maintenance activities occur like a pit stop in a NASCAR race. This insures optimum equipment uptime, with quality equipment maintenance activities being performed. Planning and scheduling pulls together all of the activities, (maintenance, operations, and engineering) and focuses them on obtaining maximum (quality) results in a minimum amount of time.

8. Is Planning and Scheduling Effective?

Although this question is similar to #6, the focus is on the efficiency and effectiveness of the activities performed in the 80% planned mode. An efficient planning and scheduling program will insure maximum productivity of the employees performing any equipment maintenance activities. Delays, such as waiting on or looking for parts, waiting on or looking for rental equipment, waiting on or looking for the equipment to be shut down, waiting on or looking for drawings, waiting on or looking for tools, will all be eliminated.

If these delays are not eliminated through planning and scheduling, then it will be impossible to optimize equipment utilization. It will be the same as a NASCAR pit crew taking too long to do a pit stop; the race is lost by not keeping the car on the track. The equipment utilization is lost by not properly keeping the equipment in service.

9. Is a CMMS / EAM System Utilized?

By this point in the Maintenance Strategy Process development, a considerable volume of data is being generated and tracked. Ultimately, the data becomes difficult to manage using manual methods. It may be necessary to computerize the work order system. If the workforce is burdened with excessive paper work and is accumulating file cabinets of equipment data that no one has time to look at, it is best to computerize the maintenance / asset management system. The systems that are used for managing the maintenance /asset management process are commonly referred to by acronyms such as CMMS (Computerized Maintenance Management Systems) or EAM (Enterprise Asset Management) systems. (The difference between the two types of systems will be thoroughly covered in Volume Four.)

The CMMS/ EAM System should be meeting the equipment management information requirements of the organization. Some of the requirements include:

• Complete tracking of all repairs and service
• The ability to develop reports, for example:
• Top ten equipment problems
  • Most costly equipment to maintain
  • Percent reactive vs. proactive maintenance
  • Cost tracking of all parts and costs

If the CMMS/EAM system does not produce this level of data, then it needs to be re-evaluated and a new one may need to be implemented.

10. Is the CMMS/ EAM System Utilization Effective?

The re-evaluation of the CMMS / EAM system may also highlight areas of weakness in the utilization of the system. This should allow for the specification of new work management process steps that will correct the problems and allow for good equipment data to be collected. Several questions for consideration include:

• Is the data we are collecting complete and accurate?
• Is the data collection effort burdening the work force?
• Do we need to change the methods we use to manage the data?

Once problems are corrected and the CMMS / EAM system is being properly utilized, then constant monitoring for problems and solutions must be put into effect.

The CMMS / EAM system is a computerized version of a manual system. There are currently over 200 commercially produced CMMS / EAM systems in the North American market. Finding the correct one may take some time, but through the use of lists, surveys, and "word of mouth," it should take no more than three to a maximum of six months for any organization to select their CMMS / EAM system. When the right CMMS / EAM system is selected, it then must be implemented. CMMS / EAM system implementation may take from three months (smaller organizations) to as long as 18 months (large organizations) to implement. Companies can spend much time and energy around the issue of CMMS selection and implementation. It must be remembered that the CMMS / EAM system is only a tool to be used in the improvement process; it is not the goal of the process. Losing sight of this fact can curtail the effectiveness of any organization's path to continuous improvement.

If the correct CMMS / EAM system is being utilized, then it makes the equipment data collection faster and easier. It should also make the analysis of the data faster and easier. The CMMS / EAM system should assist in enforcing "World Class" maintenance disciplines, such as planning and scheduling and effective stores controls. The CMMS / EAM system should provide the employees with usable data with which to make equipment management decisions. If the CMMS / EAM system is not improving these efforts, then the effective usage of the CMMS / EAM system needs to be evaluated. Some of the problems encountered with CMMS / EAM systems include:

• Failure to fully implement the CMMS
• Incomplete utilization of the CMMS

• Inaccurate data input into the CMMS
• Failure to use the data once it is in the CMMS

11. Do Maintenance Skills Training Programs Exist?

This question examines the maintenance skills training initiatives in the company. This is a critical item for any future steps because the maintenance organization is typically charged with providing training for any operations personnel that will be involved in future activities. Companies need to have an ongoing maintenance skills training program because technology changes quickly. With newer equipment (or even components) coming into plants almost daily, the skills of a maintenance workforce can be quickly dated. Some sources estimate that up to 80% of existing maintenance skills can be outdated within five years. The skills training program can utilize many resources, such as vocational schools, community colleges, or even vendor training. However, to be effective, the skills training program needs to focus on the needs of individual employees, and their needs should be tracked and validated.

12. Are the Maintenance Training Programs Effective?

This evaluation point focuses on the results of the skills training program. It deals with issues such as:
• Is there maintenance rework due to the technicians not having the skills necessary to perform the work correctly the first time?
• Is there ongoing evaluation of the employees skills versus the new technology or new equipment they are being asked to maintain or improve?
• Is there work being held back from certain employees because a manager or supervisor questions their ability to complete the work in a timely or quality manner?

If these questions uncover some weaknesses in the workforce, then it quickly shows that the maintenance skills training program is not effective. If this is the case, then a duty-task-needs analysis will highlight the content weaknesses in the current maintenance skills training program and provide areas for improvement to increase the versatility and utilization of the maintenance technicians.

13. Are Operators Involved in Maintenance Activities?

As the organization continues to make progress in the maintenance disciplines, it is time to investigate whether operator involvement is possible in some of the equipment management activities. There are many issues that need to be explored, from the types of equipment being operated, the operators-to-equipment ratios, and the skill levels of the operators, to contractual issues with the employees' union. In most cases, some level of activity is found in which the operators can be involved within their areas. If there are no obvious activities for operator involvement, then a re-evaluation of the activities will be necessary.

The activities the operators may be involved in may be basic or complex. It is partially determined by their current operational job requirements. Some of the more common tasks for operators to be involved in include, but are not necessarily limited to:

a. **Equipment Cleaning:** This may be simply wiping off their equipment when starting it up or shutting it down.

b. **Equipment Inspecting:** This may range from a visual inspection while wiping down their equipment to a maintenance inspections checklist utilized while making operational checks.

c. **Initiating Work Requests:** Operators may make out work requests for any problems (either current or developing) on their equipment. They would then pass these requests on to maintenance for entry into the work order system. Some operators will directly input work requests into a CMMS.

d. **Equipment Servicing:** This may range from simple running adjustments to lubrication of the equipment.

e. **Visual Systems:** Operators may use visual control techniques to inspect and to make it easier to determine the condition of their equipment.

Whatever the level of operator involvement, it should contribute to the improvement of the equipment effectiveness.

14. Are Operator-Performed Maintenance Tasks Effective?

Once the activities the operators are to be involved with has been determined, their skills to perform these activities need to be examined. The operators should be properly trained to perform any assigned tasks. The training should be developed in a written and visual format. Copies of the training materials should be used when the operators are trained and a copy of the materials given to the operators for their future reference. This will contribute to the commonality required for operators to be effective while performing these tasks. It should also be noted that certain regulatory organizations require documented and certified training for all employees (Lock Out Tag Out is an example).

Once the operators are trained and certified, they can begin performing their newly-assigned tasks. It is important for the operators to be coached for a short time to insure they have the full understanding of the hows and whys of the new tasks. Some companies have made this coaching effective by having the maintenance personnel assist with it. This allows for operators to receive background knowledge that they may not have gotten during the training.

15. Are Predictive Techniques Utilized?

Once the operators have begun performing some of their new tasks, some maintenance resources should be available for other activities. One area that should be explored is predictive maintenance. Some fundamental predictive maintenance techniques include:

- Vibration Analysis
- Oil Analysis
- Thermography
- Sonics

Plant equipment should be examined to see if any of these techniques will help reduce downtime and improve its service. Predictive technologies should not be utilized because they are technically advanced, but only when they contribute to improving the equipment effectiveness. The correct technology should be used to trend or solve the equipment problems encountered.

16. Are the PDM Tasks Effective?

If the proper PDM tools and techniques are used, there should be a decrease in the downtime of the equipment. Because the PDM program will find equipment wear before the manual PM techniques, the planning and scheduling of maintenance activities should also increase. In addition, some of the PM tasks that are currently being performed at the wrong interval should also be able to be adjusted. This will have a positive impact on the cost of the PM program. The increased efficiency of the maintenance workforce and the equipment should allow additional time to focus on advanced reliability techniques.

17. Are Reliability Techniques Being Utilized?

Reliability Engineering is a broad term that includes many engineering tools and techniques. Some common tools and techniques include:

a. **Life Cycle Costing:** This technique allows companies to know the cost of their equipment from when it was designed to the time of disposal.

b. **R.C.M.:** Reliability Centered Maintenance is used to track the types of maintenance activities performed equipment to insure they are correct activities to be performed.

c. **F.E.M.A.:** Failure and Effects Mode Analysis examines the way the equipment is operated and any failures incurred during the operation to find methods of eliminating or reducing the numbers of failures in the future.

d. **Early Equipment Management and Design:** This technique takes information on equipment and feeds it back into the design process to insure any new equipment is designed for maintain ability and operability.

Using these and other reliability engineering techniques improve equipment performance and reliability to insure competitiveness.

18. Are the Reliability Techniques Effective?

The proper utilization of reliability techniques will focus on eliminating repetitive failures on the equipment. While some reliability programs will also increase the efficiency of the equipment, this is usually the focus of TPM/OEE techniques. The elimination of the repetitive failures will increase the availability of the equipment. The effectiveness of the reliability techniques are measured by maximizing the uptime of the equipment.

19. Are TPM/ OEE Methodologies Being Utilized?

Are the TPM/OEE methodologies being utilized throughout the company? If they are not, then the TPM/OEE program needs to be examined for application in the company's overall strategy. If a TPM/OEE process exists, then it should be evaluated for gaps in performance or deficiencies in existing parts of the process. Once weaknesses are found, then steps should be taken to correct or improve these areas. Once the weaknesses are corrected and the goals are being achieved, then the utilization of the OEE for all equipment relate decisions is examined.

20. Is OEE Being Effectively Utilized?

The Overall Equipment Effectiveness provides a holistic look at how the equipment is utilized. If the OEE is too low, it indicates that the equipment is not performing properly and maximizing the return on investment in the equipment. Also, the upper limit for the OEE also needs to be understood. If a company were to focus on achieving the maximum OEE number, they may pay too much to ever recover the investment. If the OEE is not clearly understood, then additional training in this area must be provided. Once the OEE is clearly understood, then the focus can be switched to achieving the financial balance required to maximize a company's return on assets (ROA).

21. Does Total Cost Management Exist?

Once the equipment is correctly engineered, the next step is to understand how the equipment or process impacts the financial aspects of the company's business. Financial optimization considers all costs impacted when equipment decisions are made. For example, when calculating the timing to perform a preventive maintenance task, is the cost of lost production or downtime considered? Are wasted energy costs considered when cleaning heat exchangers or coolers? In this step, the equipment data collected by the company is examined in the context of the financial impact it has on the company's profitability. If the data exists and the information systems are in place to continue to collect the data, then financial optimization should be utilized. With this tool, equipment teams will be able to financially manage their equipment and processes.

22. Is Total Cost Management Utilized?

While financial optimization is not a new technique, most companies do not properly utilize it because they do not have the data necessary to make the technique effective. Some of the data required includes:
- MTBF (Mean Time Between Failure) for the equipment
- MTTR (Mean Time To Repair) for the equipment
- Downtime or lost production costs per hour
- A Pareto of the failure causes for the equipment
- Initial cost of the equipment

- Replacement costs for the equipment
- A complete and accurate work order history for the equipment

Without this data, financial optimization can not be properly conducted on equipment. Without the information systems in place to collect this data, a company will never have the accurate data necessary to perform financial optimization.

23. Are Continuous Improvement Techniques Utilized for Maintenance / Asset Management Decisions?

Once a certain level of proficiency is achieved in maintenance/ asset management, companies can begin to lose focus on their improvement efforts. They may even become complacent in their improvement efforts. However, there are excellent Continuous Improvement (CI) tools for examining even small problems. If new tools are constantly examined and applied to existing processes, all opportunities for improvement will be clearly identified and prioritized.

24. Are the CI Tools and Techniques Effective?

This question may appear to be subjective; however, improvements at this phase of maturity for a maintenance / reliability effort may be small and difficult to identify. However, the organizational culture of always looking for areas to improve is a true measure of the effectiveness of this step. As long as even small improvements in maintenance / reliability management are realized, this question should be answered "Yes."

25. Is Continuous Improvement Sought After in All Aspects of Maintenance / Asset Management?

When organizations reaches this stage, it will be clear that they are leaders in maintenance / reliability practices. Now, they will need continual focus on small areas of improvement. Continuous improvement means never getting complacent. It is the constant self-examination with the focus on how to become the best in the world at the company's business. Remember:

Yesterday's Excellence

is

Today's Standard

and

Tomorrow's Mediocrity

Maintenance Strategy Series Part 1

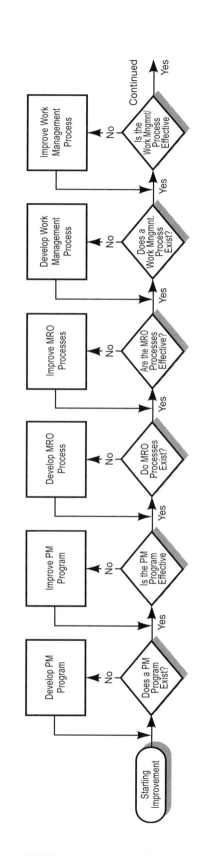

Maintenance Strategy Series Part 2

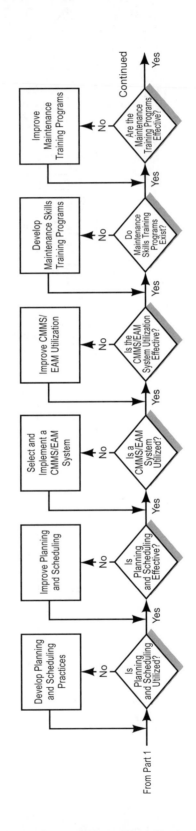

Maintenance Strategy Series Part 3

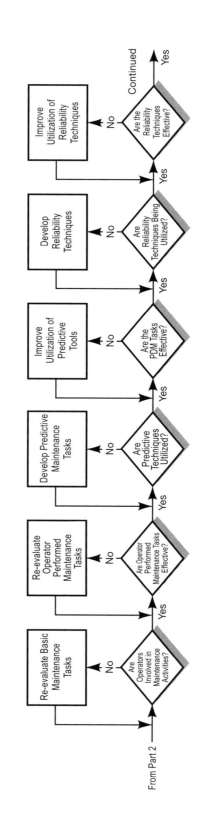

Maintenance Strategy Series Part 4

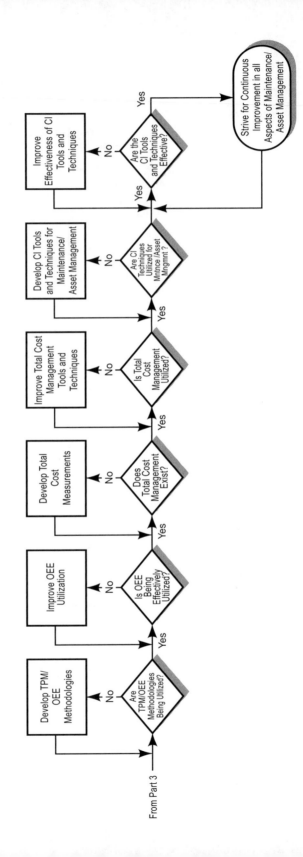

Why Preventive Maintenance is Important

If you ask twenty different people to write their definition of preventive maintenance, you will get twenty different answers. The term has varied definitions. For the purpose of this text, preventive maintenance is defined as a *fundamental, planned maintenance activity designed to improve equipment life and avoid any unplanned maintenance activity.* In its simplest form, preventive maintenance can be compared to the service schedule for an automobile. Certain tasks must be scheduled at varying frequencies, all designed to keep the automobile from experiencing any unexpected breakdowns. Preventive maintenance for industrial equipment is no different.

The Importance of Preventive Maintenance

Preventive maintenance is the foundation of the entire maintenance strategy. Unless the PM program is effective, all subsequent maintenance strategies will take longer to implement, incur higher costs to implement, and have a higher probability of failure.

This may seem to be an overstatement, but publications from authors around the world echo the same thought. Because the preventive maintenance program receives this type of solid endorsement from successful companies, it appears that companies would focus on their preventive maintenance programs, attempting to insure their success. However, this is not the case. Figure 1-1 shows the result of a survey that involved 5000 companies. As the figure highlights, the majority were not satisfied with the effectiveness of their preventive maintenance program. How is the effectiveness measured? Effectiveness occurs when 80% or more of the maintenance activities can be planned and scheduled at least one week in advance. This level is an indicator that the organization is moving from a reactive culture to a more proactive culture.

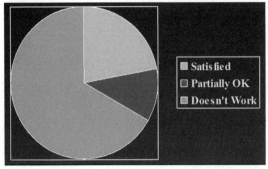

Figure 1-1
PM Program Satisfaction

The results of another survey are highlighted in Figure 1-2. This figure shows that the percentage of respondents who believe their preventive maintenance program is ineffective is almost the same as reflected in the survey in Figure 1-1. The issue is that the two surveys were taken almost 40 years apart. Figure 1-1 was taken in 1965 and Figure 1-2 was taken in 2004. Why is there a lack of progress improving the number of companies with effective preventive maintenance programs? It is due to three major reasons.

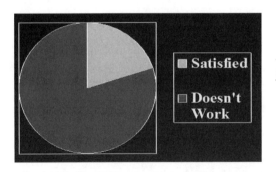

Figure 1-2
PM Program Satisfaction

The first is that companies do not focus on preventive maintenance activities because these activities are not "high profile" or "high visibility" activities. This obstacle is one of the first that companies need to overcome when moving from a reactive to a proactive culture. In a reactive culture, the "hero" technician (the individual who can fix problems quickly) is typically highly valued. The technicians then feel that the preventive maintenance activities are not valued; they will opt to focus instead on their "reactive" skills. This problem is not overcome until upper and mid level managers show they value the preventive maintenance activities more than the "fix it fast" activities. Without this management support, the technicians will always perform the activities that are perceived as the highest value.

The second issue is the pandemic originating from the lack of basic maintenance skills. In the majority of companies today, the maintenance technicians lack the skills to identify developing problems with equipment components. They are unable to perform basic lubrication tasks or even to make proper adjustments to their assigned equipment. This means that even when the organizational culture is conducive to change, the basic skills may still prevent the preventive maintenance program from being successful.

The third reason, the lack of a disciplined development process for preventive maintenance, will be discussed in Chapter 2.

Companies still try to excuse their lack of good preventive maintenance by making statements such as "Preventive Maintenance doesn't work in our industry" or "Our customers don't care about our maintenance practices." However, in every industry, there are excellent examples of companies with effective preventive maintenance programs. This calls to mind a quote from a classic textbook "Reengineering the Corporation" by Hammer and Champy. They wrote:

In almost every industry, under the same rules and with the same players, the successes of a few companies rebut the excuses of the many.

This quote is true of the maintenance/reliability business, but is especially reflective of preventive maintenance strategies. There are countless testimonials from companies that highlight the benefits of a good preventive maintenance program. The following material shows some examples from many different industries.

Concerning their equipment uptime, one discrete manufacturing company said: "We improved out equipment uptime from the 50%– 60% range to the 95% + range by instituting a preventive maintenance program."

This same company continued on about the benefits by saying "Before the routine shutdowns for preventive maintenance, we were always behind in the production schedule. After we started regular preventive maintenance shutdowns, we began to increase our production efficiencies. As a result, all operations are now shut down one shift per week for preventive maintenance or to do something to improve the process."

The magazine *Engineered Systems* presented a feature article on maintaining properly levels of humidity in a facility (a typical preventive maintenance function). The article noted, "Both people and equipment can cost companies thousands, if not millions, of dollars if the relative humidity is not maintained within the recommended guidelines." This article provides good insight into the ancillary financial contributions that an effective preventive maintenance program can have on a company's bottom line.

The magazine *Business Week* featured an article on an electronic manufacturer, where it was observed "By developing a fixed pattern for preventive maintenance chores and reinforcing them through constant repetition, the company slashed electrical breakdowns by 80% since 1990 and saved millions of dollars." This observation provides yet another example where preventive maintenance contributes to a company's profitability.

In an anecdotal story, a maintenance technician told one company's Vice-President of Operations that he considered 80–85% uptime acceptable for the plant's equipment. The Vice President changed the technician's thinking by asking "What uptime do you expect from your Chevy Blazer?" Uptime at this manufacturing company now averages 94–97%. Certainly, this was one Vice President who knew how to communicate the equipment uptime to the technicians.

The value of extending the equipment life by performing proper preventive maintenance was highlighted in an article where it was explained that "Without proper preventive maintenance, the usable life of any piece of equipment is much shorter than its design life, sometimes by as much as 30%."For a company to maximize its return on investment in an asset, it needs to have an effective preventive maintenance program. Early replacement of expensive assets will force a company to increase its capital appropriations budget. This unnecessary expense detracts from the company's profitability.

In the facilities sector, energy usage is a large portion of a company's budget. However, there are large financial impacts that an effective preventive maintenance program can have on energy expenses. For example, consider some statistics for Heating, Ventilation, and Air Conditioning:

◆ Controls for HVAC that are malfunctioning or simply out of calibration place excessive demands on equipment, causing up to a 20% greater energy demand.

◆ A condenser surface fouled with 0.015" of scale will increase energy usage by 11%.

◆ An improperly-tuned boiler will require as much as 25% more fuel to operate.

◆ A sluggish purge system on a chiller can boost energy consumption by as much as 10%.

If these items are not addressed on a proper preventive maintenance program, the energy consumption is higher than it needs to be (on average by 5% to as much as 10%) for the entire facility or plant.

In another facility example, the following parameters were given:

◆ 100,000-square-foot building
◆ 300-ton chiller, 200-hp boiler
◆ Air volume is 90,000CFM

With this established as a baseline for the size of the facility, the following typical equipment conditions were established:

◆ Refrigeration condenser are slightly scaled
◆ Coils and filters in air handlers are dirty
◆ Purge unit is not fully removing non-condensing gasses
◆ Controls is not functioning to specifications
◆ Boiler controls out of trim by 4%
◆ Cooling tower fans and nozzles are inefficient

All of these conditions are typical when a company does not have a good preventive maintenance program. The results were calculated and it was determined that the facility would be wasting approximately $25,500 per year in energy costs.

Another interesting study, published by the magazine *Preventive/ Predictive Maintenance Technology,* categorized companies into three major classifications. These were:

◆ In a breakdown mode

◆ In a preventive mode
◆ In a predictive mode

After classifying each of the organizations, the installed horsepower for each of the plants was determined. The installed horsepower was then used in the denominator to calculate the maintenance cost (the numerator) per installed horsepower. The results were:

◆ In a breakdown mode, $17–$18 per installed horsepower per year
◆ In a preventive mode, $11–$13 per installed horsepower per year
◆ In a predictive mode, $7–$9 per installed horsepower per year

This clearly shows that the more advanced a company becomes in their maintenance practices, the lower the overall maintenance costs become. While it may seem apparent that the maintenance costs (numerator) was impacted, the installed horsepower (denominator) was also lowered. The reason is the well-maintained and reliable equipment requires less redundancy. For example, the company may only need two compressors at the plant instead of three due to the reliability of the primary compressors. Furthermore, they may not need a third air compressor because the preventive maintenance program eliminates air leaks, reducing the demand form compressed air.

Finally, one other area of consideration covers the legal ramifications for companies without good preventive maintenance programs. For example, Modern HealthCare's magazine *Facility Operation* highlighted an article about hospitals with poor preventive maintenance program. In part, it noted "There are three hospitals with lawsuits filed against them, which had weak or no preventive maintenance programs at all." This statement highlights the fact that good preventive maintenance programs are essential to maintain a good standing in the community, whether a company is a hospital or any industrial type of a plant.

Additional Justification for Preventive Maintenance

Additional justifications for good preventive maintenance programs are highlighted in Figure 1-3.

Additional Preventive Maintenance Justification

- ISO, OSHA, EPA, PSM, etc.
- Total Quality Management
- Just In Time
- Customer Service Orientation
- Capacity Constraints
- Redundant Equipment
- Energy Consumption
- Usable Asset Life

Figure 1-3 Additional Preventive Maintenance Justification

Increased automation in industry requires preventive maintenance. The more automated that the equipment is, the more components there are that and fail and cause the entire piece of equipment to be taken out of service. Routine services and adjustments can keep the automated equipment in the proper condition to provide uninterrupted service.

Just-In-Time manufacturing (JIT), which is becoming more common in the United States today, requires that the materials being produced into finished goods arrive at each step of the process just in time to be processed. JIT eliminates unwanted and unnecessary inventory. However, JIT also requires high equipment availability. Equipment must be ready to operate when a production demand is made; it cannot break down during the operating cycle. Without the buffer inventories (and high costs) traditionally found in U.S. processes, preventive maintenance is necessary to prevent equipment downtime. If equipment does fail during an operational cycle, there will be delays in making the product and delivering it to the customer. In these days of intense competitiveness, delays in delivery can result in lost customers. Preventive maintenance is required so that equipment is reliable enough to develop a production schedule that, in turn, is dependable enough to give a customer firm delivery dates.

In many cases, when equipment is not reliable enough to schedule to capacity, companies will purchase another identical piece of equipment. Then if the first one breaks down on a critical order, they have a back-up. With the price of equipment today, however, this back-up can be an expensive solution to a common problem. Unexpected equipment failures can be reduced, if not almost eliminated, by a good preventive maintenance program. With equipment availability at its highest possible level, redundant equipment will not be required.

Reducing insurance inventories has an impact on maintenance and operations. Maintenance carries many spare parts in case the equipment breaks down. Operations carry additional spare parts in process inventory for the same reason. Good preventive maintenance programs allow the maintenance departments to know the condition of the equipment and prevent breakdowns. The savings from reducing (in some cases, eliminating) insurance inventories can often finance the entire preventive maintenance program.

In manufacturing and process operations, each production process is dependent on the previous process. In many manufacturing companies, these processes are divided into cells. Each cell is viewed as a separate process or operation. Furthermore, each cell is dependent on the previous cell for the necessary materials to process. An uptime of 97% might be acceptable for a stand-alone cell. But if ten cells, each with a 97% uptime, are tied together to form a manufacturing process, the total uptime for the process is only 73% (see Figure 1-4).

- **Mitigating dependencies among manufacturing cells**

Cell Number	Uptime	Total
1	97%	97%
2	97%	.97 × .97 = 94.1%
3	97%	.97 × .941 = 91.3%
4	97%	.97 × .913 = 88.6%
5	97%	.97 × .886 = 85.9%
6	97%	.97 × .859 = 83.3%
7	97%	.97 × .833 = 80.8%
8	97%	.97 × .808 = 78.4%
9	97%	.97 × .784 = 76.0%
10	97%	.97 × .760 = 73.4%
Total (10 Cells)		73.4%

Figure 1-4 Why is Preventive Maintenance Important?

This level is unacceptable in any process. Preventive maintenance must be used to raise uptime to even higher levels. Performing needed services on the equipment when required leads to longer equipment life. Returning to an earlier example, an automobile that is serviced at prescribed intervals will deliver a long and useful life. However, if it is neglected—for example, the oil is never changed—it will have a shorter useful life. Because industrial equipment is often even more complex than the newer computerized automobiles, service requirements may be extensive and critical. Preventive maintenance programs allow these requirements to be met, reducing the amount of emergency or breakdown work the maintenance organization is required to perform.

Preventive maintenance reduces the energy consumption for the equipment to its lowest possible level. Well-serviced equipment requires less energy to operate because all bearings, mechanical drives, and shaft alignment receive timely attention. By reducing these drains on the energy used by a piece of equipment, overall energy usage in a plant can decrease by 5% to as much as 11%.

Quality is another cost reduction that helps justify a good preventive maintenance program. Higher product quality is a direct result of a good preventive maintenance program. Poor, out-of-tolerance equipment never produces a quality product. World class manufacturing experts recognize that rigid, disciplined preventive maintenance programs produce high quality products. To achieve the quality required to compete in the world markets today, preventive maintenance programs are required.

If operations or facilities were organized and operated the way the majority of maintenance organizations are, we would never get any products or services when we needed them. An attitude change is necessary to give maintenance the priority it needs. This change also includes management's viewpoint. Modern management tends to sacrifice long-term planning for short-term returns. This attitude causes problems for maintenance organizations, leading to reactive maintenance with few or no controls. When maintenance is given its due attention, it can become a profit center, producing positive, bottom line improvements to the company.

No preventive maintenance program will be truly successful without strong support from the facility's upper management. Many decisions must be made by plant management to allow time to perform maintenance on the equipment instead of running it wide open. Without upper management's commitment to the program, PM will either never be performed, or it will be performed too little, too late. Thus, management support is the cornerstone for any successful PM program.

2

DEVELOPING THE PREVENTIVE MAINTENANCE PROGRAM

As discussed in the last chapter, there are many benefits from developing, implementing, and properly executing a preventive maintenance program. Yet, as also pointed out in the last chapter, the success rate for companies implementing preventive maintenance programs is very small.

Two of the major reasons why preventive maintenance programs fail were discussed in Chapter 1. The third and actually the primary reason is the lack of a disciplined methodology to implement a preventive maintenance program. One of the major contributing factors to this is the failure to visualize the preventive maintenance process.

If the business flow for a preventive maintenance program is visualized, it gives the organization an outline to follow. A typical flow for a preventive maintenance process is pictured in Figure 2-1. The following material will briefly outline the preventive maintenance process. Each major topic area on the outline will be detailed in the following chapters.

Developing or Changing the Preventive Maintenance Program

When the decision is made to develop or change the preventive maintenance program, there is usually an immediate business reason to do so. If there is new equipment being installed, then someone has developed the business case to justify the purchase of the new equipment. This justification should have also included the life cycle costs, which should include the cost of performing the preventive maintenance throughout the life of the equipment.

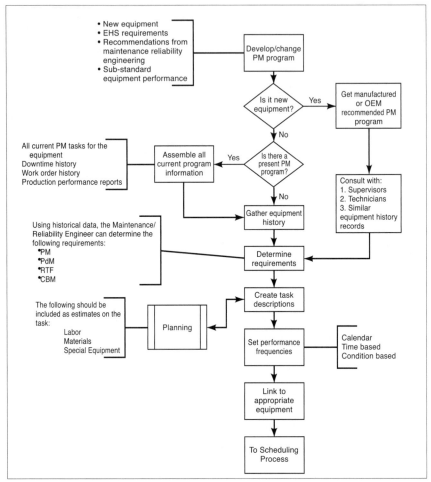

Figure 2-1 Preventive Maintenance Program Management

Is it New Equipment?

If the equipment is new, a different track is taken than if the equipment is existing. If the equipment is new, the following steps should be followed:

◆ Obtain the manufacturer or OEM-recommended PM program

In this step, you simply request the recommended maintenance tasks and service intervals from the manufacturer of the equipment. While this does not comprise the entire PM program for the equipment, it forms the foundation.

◆ Consult with:
   1. Supervisors
   2. Technicians
   3. History records of similar equipment

Once the manufacturer's PM program is obtained, it is further refined by reviewing it with the existing maintenance supervisors, maintenance technicians (including the planners) and similar equipment history records. This will allow for refinement of the program before it is initiated on the new equipment.

If the PM program is being developed or changed for existing equipment, the process moves to:

Is There a Present PM Program?
This step is used to determine if there are existing preventive maintenance inspections and tasks. If inspections and tasks already exist, then they should be examined for areas of improvements. If there is no existing preventive maintenance program, then you must gather existing data to develop the proper preventive maintenance program.

◆ **Assemble all current program information**
If there is an existing preventive maintenance program, it is important to collect all of the existing information related to the program. This will include:
   1. All of the existing tasks—to check for proper amount of detail and coverage of the equipment

   2. Downtime History—to evaluate the effectiveness of the existing PMs

   3. Work Order History—to evaluate the amount of work that is requested on the equipment not currently found by the PM program

   4. Production Performance Reports—to evaluate how close to scheduled performance the equipment is achieving

   5. Design Engineering Specifications—to evaluate how close to actual design performance the equipment is capable of achieving

Gather Equipment History

In this step, we compile all of the previous information into a historical record. Doing so provides the detail needed to plot the entire life cycle of the equipment. The compilation of this data insures that it is clearly understood where the equipment is in its life cycle so that the proper PM tasks can be specified.

Determine the Requirements

Using the data that has been gathered to this point, the maintenance or reliability engineer and the maintenance planner for the equipment can begin developing the preventive maintenance program. Using the data, the proper blend of the following strategies can be applied:

1. PMs such as inspections, routine services, lubrication, and basic adjustments

2. PdM techniques, such as vibration analysis, oil analysis, thermography, and sonics

3. Condition-based techniques, which are usually real time tools that use PdM technologies mentioned in the previous strategy

4. RTF (Run To Failure) strategies, which are used when the previous options are not feasible. The failure usually will not endanger the safety of employees, cause a production delay, or cause a significant financial impact on the company.

Once the various strategies are determined for the equipment, the task development can begin.

Create the Task Descriptions

In this step, the actual work to be performed is detailed. This step includes developing a complete job plan (hence the reference to planning) for the PM task. Therefore, the planner will be required to detail the skilled labor requirements, the spare parts required, and any special tools or equipment that will be required. Once the plan is finalized, the time estimation should be made for the PM task.

Set Performance Frequencies

This step sets the frequency for performing the PM. Typical factors for determining when the PM will be performed are:
1. Calendar based
2. Usage based
3. Condition based

Once the frequency is set, the PMs are ready to be attached to the equipment.

Link to the Appropriate Equipment

While the PMs are developed with a specific equipment focus, most plants have similar, if not identical equipment. At this step, the PMs are attached to each piece of equipment. Developing the PMs in this manner allows for rapid development of a complete PM program with the least amount of resources.

To the Scheduling Process

In this step, the schedule for the PM program should be balanced. This means scheduling the PM program so that as the PMs come due they require a balanced amount of resources for each scheduling period. For example, it would be unwise to schedule 100 hours of PMs one week and 10 the next. An unbalanced load will make scheduling difficult and typically will have a negative impact on PM compliance percentages.

In this chapter, we explained the general flow of the PM process. In the following chapters, we will explore each of the process steps in further detail.

3

IDENTIFYING THE EQUIPMENT TO INCLUDE IN A PREVENTIVE MAINTENANCE PROGRAM

Develop the Critical Equipment Units and Systems List

The first step of designing a PM program is to determine the critical units and systems in the plant that will be included in the PM program. Maintenance managers know that having the PM program cover every item in the plant or facility is not cost effective. There are certain components, not part of critical processes, which are cheaper to let run to failure than to spend money maintaining. Critical items should be identified and cataloged for inclusion in the PM program.

Determining the critical equipment items can be accomplished several ways. It can be by:

◆ The highest amount of downtime
◆ The highest lost production costs
◆ The biggest quality problems

Regardless of the method used to determine the equipment on which the PM program is started, the primary concern is that the equipment is considered to be a bottleneck or a constraint to producing a product.

If a line or process is capacity constrained, then any downtime by equipment in the line or process has a direct impact on the product throughput. Even if the product cannot be sold (a capped market), the overhead cost to produce the product will be higher than necessary.

If a line or process is involved in producing a product that is in a demand marketplace—and can be sold immediately upon being produced—then any downtime become lost revenue. This type of condition is ideal for cost justifying a preventive maintenance program.

If a line or process has quality problems (a reject or rework percentage greater than 1%), then the process should be examined to see if the equipment is causing the quality problem. In some case studies, up to 50% of all quality problems are related to equipment-induced problems. If the quality problems can be traced to equipment issues, then preventive maintenance tasks should be able to reduce or eliminate the defects.

One of the major manufacturing strategies, the *Theory of Constraints,* is a perfect match to a preventive maintenance program. This strategy highlights a problem piece of equipment in a line or process that restricts the manufacturing throughput. If the constraint equipment has performance problems related to maintenance, then the preventive maintenance program plays a key role in removing the constraint. If the preventive maintenance program is to be effective, it must be focused on the equipment problems causing the equipment not to perform to design specifications.

Another consideration for classifying equipment as critical is if it falls under some regulatory program, such as the Environmental Protection Agency (EPA), the Food and Drug Administration (FDA), or even if it is critical to a company's International Standards Organization (ISO) certification.

Prioritization of the Critical Equipment List

Once the critical equipment has been identified, regardless of the method used, the equipment list should be prioritized. The prioritization should follow the critical equipment order. The second method would be to sort the critical equipment by its impact on the production or operations. This would include the lost production due to maintenance-related problems. The problems could result in either equipment downtime or with the equipment performing at less than design capacity.

The last step, which most companies do not take, is to quantify the valuation of the lost production. This step will eventually be utilized to financially justify the preventive maintenance program.

Age of the Equipment

With the critical equipment identified and prioritized, the next step will be to identify each individual piece of critical equipment as to whether it is new equipment or existing equipment. New equipment will take a slightly different developmental path than existing equipment. Because the company will not have any history with the new equipment, it will have to rely heavily on the manufacturer's recommended maintenance intervals.

If the critical equipment is existing equipment, then the preventive maintenance program development relies more heavily on the historical equipment performance and the detailed failure history. This insures that existing problems with the equipment will be properly addressed by the preventive maintenance program.

Simplicity versus Complexity

Once the sources for the preventive maintenance materials have been clarified, it is time to begin breaking the equipment down into its component parts. This process involves what is commonly called an equipment tree. In other cases, this may be referred to as the Bill of Material for a piece of critical equipment. (The latter phrase is more commonly encountered when establishing a spare parts list for a piece of equipment.) The goal of utilizing this process is to divide the equipment into more common components so the preventive maintenance procedures will be easier to develop. For example, if one were to examine a Cold Mill (in a steel mill), it might seem overwhelming in complexity. However, if the mill is broken down into component systems, it is easier to visualize how to maintain it. It will have multiples of the following:

◆ Electrical Systems
◆ Mechanical Systems
◆ Fluid Power Systems

Each of these component systems can be further divided into:
◆ Electrical Systems
High Voltage (440 Volts and above)
Operational Voltage (120 Volts to 440 Volts)
Control Voltage (Voltages less than 120 Volts)

◆ Mechanical Systems
    Belt Drives
        V-Belts
        Synchronous Belts
        Flat Belts
    Chain Drives
        Roller Chain
    Bearings
        Roller Bearings
        Ball Bearings
        Sleeve Bearings
    Gear Drives
        Spur Gears
        Helical Gears
        Worn Gears
    Couplings
        Rigid Couplings
        Flexible Couplings

◆ Fluid Power Systems
        Hydraulics
        Pneumatics

While each of these components can be further sub-divided, this level is sufficient to make the point. For example, what type of preventive maintenance procedures are required for a V-belt drive? It is not to difficult to determine the correct preventive maintenance procedures. (Chapter 5 will deal with this topic extensively.)

Figure 3-1 shows a typical hierarchy. Using the previous example of a cold mill, the mill can be broken down into three major sections: the entry end, the mill stand, and the exit end. Focusing on the entry end, it has three major components: the hydraulic system, the payoff reel, and the coil buggy. Again, focusing only on the hydraulic system, it has a pump, a motor, and a control system. (Note that the list of components is truncated to simplify the example)

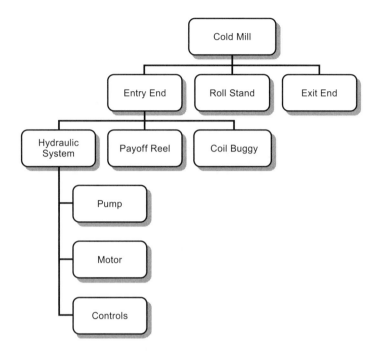

Figure 3-1 Equipment Component Hierarchy

With this level of detail, it becomes easier to ask:

◆ What preventive maintenance services does the pump require?
◆ What preventive maintenance services does the motor require?
◆ What preventive maintenance services does the control system require?

Defining the preventive maintenance system at the component level provides a comprehensive listing of the services required. Using this component listing, it is then easy to combine the lists for the equipment, whether it is broken out by a functional level (entry section) or by a system level (electrical system).

Figure 3-2 shows a functional hierarchy for a power plant. It begins with multiple systems and divides them into their basic components, similar to the cold mill. This diagram makes use of the system level division. The level used (whether functional or system) does not really matter. What does matter is choosing one and staying consistent with its usage throughout the preventive maintenance program development.

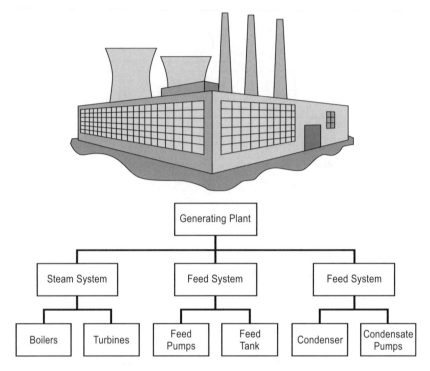

Figure 3-2 Equipment Hierarchy

This chapter has focused on determining the critical equipment on which to begin the development of the preventive maintenance program. It has also shown how to divide complex equipment systems into more maintainable components. The next chapter will explore the various sources for developing good detailed preventive maintenance procedures.

How to Develop PM
Requirements for the Equipment

This step will develop specific types of services that are required for each component, including all the techniques that apply to the components that were described in the previous section. These general descriptions of the services could serve as titles to the checklists and procedures that will be developed in the next step.

The information for the general lists of services that must be performed on each equipment item can be found in four basic areas listed in Figure 4-1.

Sources of Preventive
Maintenance Information

- Manufacturers
- Maintenance Engineers
- Operators, Craft Technicians, Supervisors
- Regulatory Agencies

Figure 4-1 Sources of Preventive Maintenance Information

All four should be used and the results compared to insure full coverage for the equipment components. Each area has its strengths and weaknesses. The manufacturers will tend to over-maintain the components. The maintenance engineers use the past to predict the future. Any changes in equipment design, operating levels, or maintenance tools and techniques can alter the component's maintenance needs. Operators, craft technicians, and supervisors all tend to rely on their memories and can forget or overlook items. If all three are combined, the correct maintenance services can be listed.

Manufacturer Recommendations

When equipment is manufactured, the maintenance requirements should be specified by the manufacturer. In these cases, the maintenance requirements are determined by the design of the equipment. Because the equipment is designed and produced by the manufacturer, it is only reasonable that the equipment manufacturer should specify the maintenance requirements.

The release of the maintenance instructions to the company purchasing the equipment should be specified during the purchasing cycle for the equipment. This detail seems to be common sense. However, in many cases, the equipment is purchased without consultation with the maintenance department, so the maintenance details are often overlooked. When the equipment is purchased, the following should be provided by the manufacturer:

◆ Operation manuals
◆ Maintenance manuals
◆ Recommended spare parts lists
◆ Preventive maintenance recommendations

Several areas need to be considered when reviewing the manufacturer's preventive maintenance recommendations. The first is the level at which the information is written. Is the material written a level at which your plant technicians can comprehend? Equipment manufacturers use a variety of individuals to write their preventive maintenance instructions. Depending on whether the individual is an engineer, someone with a maintenance background, a technical writer, or a clerical person, the detail and usefulness of the instructions will vary. If the instructions are written at the proper level, the plant personnel will be able to quickly add the preventive maintenance procedures and schedules of the new equipment to the database that is currently being utilized for the existing plant equipment.

A lesson could be learned by equipment manufacturers from the automotive industry. Every automobile produced comes with an owners manual. In the owners manual there are sections covering familiarization of the automobile, operating instructions, maintenance instructions (including recommended schedules), and minimum standards for the lubricants and other fluids required by the automobile. This information should be provided to and kept by the purchaser of any industrial equipment.

A second point that could be gleaned from the automobile manufacturers—the reading level at which the owners manual is written. The reading level in U.S. plants averages the 8th- grade level. Therefore, providing manuals written at this level would greatly improve the acceptance of the new equipment and insure that preventive maintenance and operating instructions are properly implemented.

The major disadvantage to strictly implementing the manufacturer's recommendations is they want the equipment to last as long as possible. In trying to achieve this, they will tend to specify over-maintaining the equipment. To prevent this, companies will need to balance the manufacturer's preventive maintenance recommendations with the company's experience with similar equipment. Similarly, some companies will over-maintain their equipment trying to get the maximum life out of it. However, with the focus on cost reductions prevalent in so many plants today, the companies that spend too much on maintenance of their equipment are the exceptions, rather than the rule.

Maintenance Engineers

The maintenance engineer in most companies has responsibility for the preventive maintenance program. If the responsibility for the preventive maintenance program falls to someone else, the maintenance engineer will be a major resource for this individual.

Why the maintenance engineer? The maintenance engineer typically will have all or some of the following responsibilities:

- Insures that equipment is properly designed, selected, and installed based on life-cycle philosophy.
- Insures that equipment is performing effectively and efficiently.
- Establishes and monitors programs for engine/compressor analysis and vibration and other condition-monitoring techniques.
- Reviews deficiencies noted during corrective maintenance.
- Provides technical guidance for CMMS.

- Maintains and advises on use and disposition of stock, surplus, and rental rotating equipment.
- Promotes equipment standardization, recommends spare parts levels, and coordinates sharing of spare parts with other asset teams.
- Available for consultation with maintenance technicians.
- Monitors new technology and keeps management/staff appraised on the new developments.
- Champions quality assurance services including shop qualifications for outside services.
- Develops standards and procedures for major maintenance jobs.
- Periodically makes cost/benefit review of maintenance management programs for areas of responsibility and exchanges information across asset teams.
- Provides technical guidance for PM and PDM programs.
- Monitors competitors activities in the field of Maintenance Management.
- Focal point for monitoring performance indicators for maintenance management program.
- Optimizes maintenance strategies.
- Focal point for analyzing equipment operating data.

As these responsibilities are examined, it is clear that the role of maintenance engineers is asset based, not crew based. In other words, maintenance engineers will be involved in any technical activity (including preventive maintenance) that affects asset performance.

This implies that maintenance engineers will constantly work on improving the asset performance. Preventive maintenance activities, like inspections and services, will extend equipment life and reduce unplanned downtime. Thus, maintenance engineers will be keenly interested in how the program is structured.

Maintenance engineers will use the equipment history, manufacturer's recommendations, and input from both the planners and other maintenance personnel to develop and maintain an effective preventive maintenance program for the critical plant equipment. Again, if maintenance engineers are not the individuals responsible for the preventive maintenance program, they should be a major resource to those individual who have the primary responsibility.

Operators, Craft Technicians, Supervisors

When developing the preventive maintenance program, this group of individuals in the plant can provide insight into how the equipment is currently performing. The information derived from this group is usually not documented in the company's CMMS (Computerized Maintenance Management System) or EAM (Enterprise Asset Management) system.

Operators

The operators run the equipment each day they are at work. They come to recognize any changes in the equipment, such as an increase in noise, vibration, temperature, etc. Although they do not always know why the changes are occurring, they can alert the appropriate maintenance personnel to the change. Ultimately, when the root cause of the problem is discovered, the operators will remember the cause. Then, the next time a change is observed, the operators will remember the cause. When consulting them about the equipment, these causes can be highlighted and then become the focus of the services and inspections as a part of the preventive maintenance program. This information can be captured on daily checklists designed specifically for the operators. The maintenance engineer can be invaluable in designing the content for the checklists so it can be recorded and analyzed. Some sample operator checklists can be found in Appendix A.

Craft Technicians

The craft technicians perform maintenance on the equipment on an almost daily basis. Like the operators, they become familiar with the undocumented nuances of the equipment. They will make observations similar to that of the operators, but with more of a technical focus. The craft technicians are more likely to see through to the root cause of the problem than are the operators. The preventive maintenance inspections and services for the craft technicians should be more technical and specific in nature. This information should be captured in a format that will easily be entered into the CMMS or EAM system (in the equipment history) for further analysis by the maintenance engineer. Some sample craft technician preventive maintenance inspections can be found in Appendix B.

Supervisors

The supervisors, whether from operations or maintenance, are more focused on the equipment performance and will likely recognize higher-level equipment performance issues. They are not likely to perform equipment inspections as are the operators or craft technicians. However, they may perform area inspections for cleanliness and safety-related issues. When walking down the equipment, they may see items that should be repaired, replaced, or adjusted. These should be noted and submitted as refinements to the preventive maintenance program.

Regulatory Agencies

In the United States (as with other countries as well) there are various regulatory agencies that monitor the maintenance of a company's equipment. For example, OSHA (Occupational Health and Safety Administration) focuses on proper equipment guarding, scaffolding, work practices, etc. The EPA (Environmental Protection Agency) is concerned that equipment is maintained to a level where it does not endanger the environment. The FDA (Food and Drug Administration) works to insure that good maintenance practices are in place to safeguard the food and drug supplies. For example, in the FDA's guidelines, the following is stated:

The firm should maintain a documented maintenance program for all equipment. Maintenance should be performed on at least the minimum schedule recommended by the equipment manufacturer.

If a company is responsible for obeying the FDA's guidelines, then their preventive maintenance program will be heavily influenced by the organization's regulatory standards. In past cases where the guidelines have been disregarded, the regulatory agencies have levied large fines for non-compliance.

In a similar document from a regulatory agency in Germany, the following is observed:

The maintenance plan includes the entirety of all maintenance work operations. It makes reference to the instructions for action relative to the individual maintenance and inspection operations. These, in turn, include information about:

- Measuring variables, test, and calibration data including allowed deviations
- Necessary measuring, inspection, and calibration equipment
- Necessary special tools
- Attachment points

- Weights
- Required utilities and auxiliaries
- Reference to particular hazard points (such as corrosive liquids, components subjected to temperature, pressure or tension, etc.)
- Safety equipment and safety measures
- Required personal protective equipment
- Supplementary documentation (e.g., circuit diagrams, explosion-view drawings, inspection and maintenance location overview, lubrication instructions, list of wearing components, etc.)

These two examples highlight the need for paying careful attention to the regulatory requirements when developing the preventive maintenance program. Otherwise, a company could find themselves in a serious situation.

This point was highlighted in a 2006 benchmarking survey of over 800 companies. In this survey, 17% of the respondents replied they did not have sufficient maintenance resources to execute the preventive maintenance tasks necessary to keep their company's assets in compliance with regulatory agencies.

Understanding the resources when preventive maintenance information can be gathered is important if a company is to develop a comprehensive program. It is also important if a company is to develop the proper level of details for their preventive maintenance plans. The next chapter will discuss the development of these plans.

5

HOW TO DEVELOP PM TASK SHEETS

Types of Preventive Maintenance

The various types of preventive maintenance are listed in Figure 5-1. A good PM program will incorporate all of these types, with the emphasis varying from industry to industry and from facility to facility. This list also provides a progressive methodology for implementing a progressive preventive maintenance program.

- **Basics**
  - Cleaning, Inspecting, Lubrication, Fastening
- **Proactive Repairs or Replacements**
- **Shutdowns, Outages, or Overhauls**
- **Predictive Maintenance**
  - Vibration, Oil Analysis, Thermography, Sonics
- **Condition Monitoring**
- **Reliability Centered Maintenance**

Figure 5-1 Types of Preventive Maintenance

Basic Preventive Maintenance

Basic preventive maintenance—including lubrication, cleaning, and inspections—is the first step in beginning a preventive maintenance program. These service steps take care of small problems before they cause equipment outages. The inspections may reveal deterioration, which can be repaired through the normal planned and scheduled work order system. One problem develops in companies that have this type of program: they stop here, thinking this constitutes a preventive maintenance program. However, it is only a start; a company can do more.

Proactive Replacements

Proactive replacements substitute new components for deteriorating or defective components before they can fail. This repair schedule eliminates the high costs related to a breakdown. These components are usually found during the inspection or routine service. One caution: Replacement should be only for components in danger of failure. Excessive replacement of components thought but not known to be defective can inflate the cost of the preventive maintenance program. Only components identified as defective or "soon to fail" should be changed.

Scheduled Refurbishings

Scheduled refurbishings are generally found in utility companies, continuous process-type industries, or cyclic facilities, such as colleges or school systems. During the shutdown or outage, all known or suspected defective components are changed out. The equipment or facility is restored to a condition where it should operate relatively trouble free until the next outage. These projects are scheduled using a project management type of software, allowing the company to have a time line for starting and completing the entire project. All resource needs are known in advance, with the entire project being planned.

Predictive Maintenance

Predictive maintenance is a more advanced form of the inspections performed in the first part of this section. Using the technology presently available, inspections can be performed that detail the condition of virtually any component of a piece of equipment. Some of the technologies include:

◆ Vibration analysis
◆ Spectrographic oil analysis

♦ Infrared scanning
♦ Shock pulse method

The main differentiation between preventive and predictive maintenance is that preventive maintenance is more of a basic task, whereas predictive maintenance uses some form of a technology. (This topic will be covered in extensive detail in Volume 7 of the Maintenance Strategy Series.)

Condition-Based Maintenance

Condition-based maintenance takes predictive maintenance one step further, by performing the inspections in a real-time mode. Sensors installed on the equipment provide signals that are fed into the computer system, whether it is a process control system or a building automation system. The computer then monitors and trends the information, allowing maintenance to be scheduled when it is needed. This eliminates error on the part of the technicians who would otherwise make the readings out in the field. The trending is useful for scheduling the repairs at times when production is not using the equipment. (This topic will also be covered in extensive detail in Volume 7 of the Maintenance Strategy Series.)

Reliability Engineering

Reliability engineering, the final step in preventive maintenance, involves engineering. If problems with equipment failures still persist after using the aforementioned tools and techniques, engineering should begin a study of the total maintenance plan to see if anything is being neglected or overlooked. If not, a design engineering study should be undertaken to study possible modifications to the equipment to correct the problem. Incorporating all of the above techniques into a comprehensive preventive maintenance program will enable a plant or facility to optimize the resources dedicated to the PM program. Neglecting any of the above areas can result in a PM program that is not cost effective. (This topic will be covered in extensive detail in Volume 8 of the Maintenance Strategy Series.)

Develop Detailed Procedures and Job Plans

At this point, decisions have been made as to the critical equipment.

The critical equipment has been sub-divided into its various components. The types of preventive maintenance that will be most effective for the equipment have been determined. Finally, the sources of information about the service requirements for the equipment have been identified.

One of the first decisions that needs to be made covers the types of preventive maintenance tasks that are being developed. For example, are they to be divided into inspections, basic servicing, disassembly inspections, or proactive replacements? Some organizations will choose to make these all separate preventive maintenance activities, whereas others will combine the tasks into one activity.

What is the determining factor in the decision? It is related to who will execute the preventive maintenance tasks. If the tasks will eventually be performed by operators, then it is usually better to separate out the activities. If the tasks will always be performed by a senior level maintenance technician, then the tasks can be combined.

With this decision made, the next step is to begin developing the detailed job plans (procedures) that will be required to carry out the preventive maintenance. These job plans should include the following information:

- Detailed job instructions, including safety directions
- The required craft
- The amount of time required for the craft
- Any shutdown or downtime requirements
- A listing of all materials required

The first step in developing the preventive maintenance procedures is the detailing of the steps necessary to properly complete the task. Unfortunately, the majority of companies fail to properly detail their preventive maintenance tasks. They use vague or subjective terms and instructions that leaves the actual work up to the supervisors' or the craft technicians' interpretation of the instructions. Consider several of the statements in Figure 5-2. When a preventive maintenance task leaves the instructions vague, it leads to mistakes. Try, for example, the first line. When would a craft technician consider a motor to be hot? Would it likely depend on the technician's background and expertise? More experienced technicians might recognize a problem, whereas these with lesser skills would not recognize it.

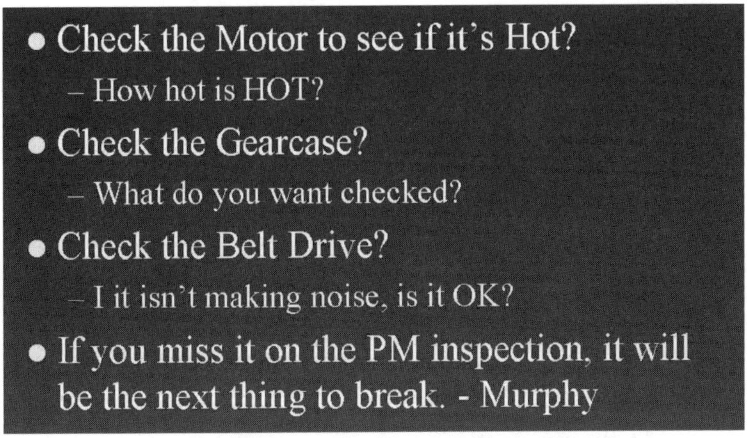

- Check the Motor to see if it's Hot?
  - How hot is HOT?
- Check the Gearcase?
  - What do you want checked?
- Check the Belt Drive?
  - I it isn't making noise, is it OK?
- If you miss it on the PM inspection, it will be the next thing to break. - Murphy

Figure 5-2 Preventive Maintenance Details

Training and Preventive Maintenance

Another major opportunity that goes undiscovered for most companies is the potential training that can occur when the preventive maintenance tasks are properly detailed. This opportunity is highlighted in Figure 5-3. If the preventive maintenance tasks are properly detailed, they can become a document by which to train the apprentice craft technicians. As time progresses, by the end of this decade, attrition will begin removing the majority of the skilled technicians that are part of the "Baby Boomer" generation from the work force. The personnel with the skills to replace them simply do not exist. This does not apply strictly to one country or continent, but is a world-wide problem.

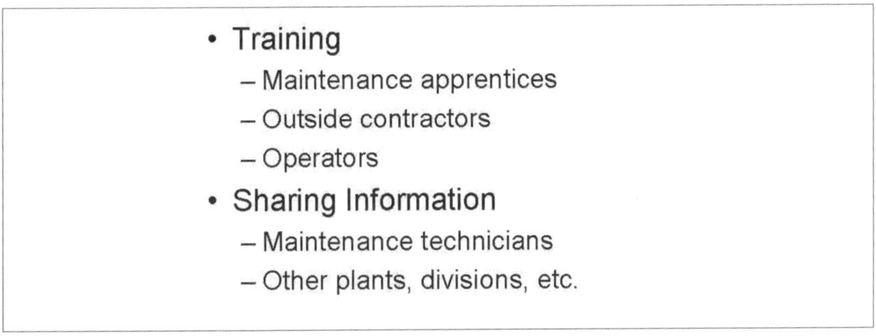

- Training
  - Maintenance apprentices
  - Outside contractors
  - Operators
- Sharing Information
  - Maintenance technicians
  - Other plants, divisions, etc.

Figure 5-3 Hidden Preventive Maintenance Opportunities

If the knowledge that will be exiting the work force over the next few years can be captured on detailed preventive maintenance and work plans, then the documents to develop the skills training to allow industry to survive will be preserved. In fact, this level of information allows cost-effective skills training that can specifically focus on the equipment in a particular plant or process. Unfortunately, this level of detail will require some time from the more experienced craft technicians—NOW—before they retire and leave the work force.

An Operating Preventive Maintenance Task

What does a typical preventive maintenance task description look like? Figure 5-4 shows an actual preventive maintenance task taking from a company in the food industry. Remember, though, this is a typical level of detail found in most plants today. Using this example, follow the evo-

WEEKLY PM OPERATING MEAT APPLICATOR

INITIAL EACH ITEM UPON COMPLETION

MEAT APPLICATOR

(___) 1. Check product belt for condition.
(___) 2. Check scraper blade for tension & effectiveness.
(___) 3. Check meat spreader system.
(___) 4. Check condition of bearings & clevis.
(___) 5. Make sure cylinder isn't leaking.
(___) 6. Make sure air lines are hooked up &operational.
(___) 7. Check carousel nylon disc for wear.
      Check amount of vibration which could be sign of worn out bushings.
(___) 8. Check that all contacts are being made on guard covers and clamped down.
(___) 9. Check all electrical conduits, cords, plugs & panels for broken connectors
      and faulty wiring.
(___) 10. Visually look at bearing for wear.
(___) 11. Lubricate bearings.
(___) 12. Check Product Zone for loose or peeling paint on motors, gearboxes, supports and
      walls and ceilings.
(___) 13. Check Product Zone for loose or missing caulk. Check Product Zone for rusty motors,
      gearboxes, supports and walls and ceilings. (Product Zone is any area above where
      product is exposed.)

Figure 5-4 Weekly PM Operating Meat Applicator

WEEKLY OPERATING MEAT APPLICATOR

INITIAL EACH ITEM UPON COMPLETION

MEAT APPLICATOR

(____)   1. Check product belt for condition.
          a. Write a PM Follow-up notification for any repairs or replacements that are required

(____)   2. Check scraper blade for tension & effectiveness.
          a. Write a PM Follow-up notification for any repairs or replacements that are required

(____)   3. Check meat spreader system.
          a. Write a PM Follow-up notification for any repairs or replacements that are required

(____)   4. Check condition of bearings & clevis.
          a. Write a PM Follow-up notification for any repairs or replacements that are required

(____)   5. Make sure cylinder isn't leaking.
          a. Write a PM Follow-up notification for any repairs or replacements that are required

(____)   6. Make sure air lines are hooked up & operational.
          a. Write a PM Follow-up notification for any repairs or replacements that are required

(____)   7. Check carousel nylon disc for wear.
          a. Write a PM Follow-up notification for any repairs or replacements that are required

(____)   8. Check that all contacts are being made on guard covers and clamped down.
          a. Write a PM Follow-up notification for any repairs or replacements that are required

(____)   9. Check all electrical components

(____)   10. Visually look at bearing for wear.

(____)   11. Lubricate bearings.

(____)   12. Check Product Zone for loose or peeling paint on motors, gearboxes, supports and walls and ceilings. (Product Zone is any area above where product is exposed.)

(____)   13. Check Product Zone for loose or missing caulk. (Product Zone is any area above where product is exposed.)

(____)   14. Check Product Zone for rusty motors, gearboxes, supports and walls and ceilings. (Product Zone is any area above where product is exposed.)

Figure 5-5 Weekly Operating Meat Applicator, 2nd Attempt

lution of details from a vague and obscure preventive maintenance task to one that can be used as a detailed task that will be useful on many levels.

Notice this is a weekly preventive maintenance task that is carried out while the equipment is operating. Consider the first task. If one were to give this to anyone but the most experienced craft technician, what would they consider an acceptable condition to be? In line 2, how much tension is appropriate? In line 3, what is to be the check on the meat spreader system? What would be acceptable and what would require additional service or work?

PMP-504 WK OPERATING MEAT APPLICATOR
INITIAL EACH ITEM UPON COMPLETION
KYA00147 MEAT APPLICATOR

(____) 1. Check product belt for condition.
    a. **What are the conditions that are objectionable enough to note as needing repair/ replacement?**
    b. Write a PM Follow-up notification for any repairs or replacements required.

(____) 2. Check scraper blade for tension & effectiveness.
    a. **What is the proper amount of tension?**
    b. **How does the inspector determine if the blade is effective?**
    c. Write a PM Follow-up notification for any repairs or replacements required.

(____) 3. Check meat spreader system.
    a. **What are items that should be listed for checking?**
    b. Write a PM Follow-up notification for any repairs or replacements required.

(____) 4. Check condition of bearings & clevis.
    a. **What is an acceptable condition of the bearings and the clevis?**
    b. Write a PM Follow-up notification for any repairs or replacements required.

(____) 5. Make sure cylinder isn't leaking.
    a. Listen for air leaks.
    b. Look for loose or damaged lines.
    c. Check the rod end of the cylinder for blow through.
    d. Write a PM Follow-up notification for any repairs or replacements required.

(____) 6. Make sure air lines are hooked up &operational.
    a. Look for loose of damaged lines.
    b. Check for loose fittings.
    c. Listen for any leaks.
    d. Write a PM Follow-up notification for any repairs or replacements required.

Figure 5-6 Weekly Operating Meat Applicator, Final Changes

Just with this example, one could quickly see the need for improvement in the detail. Because this is an operating preventive maintenance task, the lock out tag out instructions would not be included. How could this preventive maintenance procedure be improved? Figure 5-5 shows a second attempt at the detail after some training for the planners occurred.

While this level of detail appears to be improved, there was little change, except for the addition of asking for follow up work orders to be written. The preventive maintenance task appears to be longer, but is still insufficient in detail. After some additional training and clarification, a

---

(___) 7. Check carousel nylon disc for wear.
      a. Observe for any vibration, which could be sign of worn out bushings.
      b. Write a PM Follow-up notification for any repairs or replacements required.

(___) 8. Check that all contacts are being made on guard covers and clamped down.
      a. Write a PM Follow-up notification for any repairs or replacements required.

(___) 9. Check all electrical components for:
      a. Conduit covers all in place
      b. Damage to conduit
         i. bent or worn sections
      c. All cords and plugs for
         i. Worn or frayed areas
         ii. Damaged connectors
         iii. Taped areas
      d. Write a PM Follow-up notification for any repairs or replacementsrequired.

(___) 10. Visually look at bearing for wear.
      a. Excessive play:
         i. if there is visible movement in the races or between the inner race and shaft, write a PM Follow-up notification for replacement.
(___) 11. Lubricate bearings.
      a. Grease the bearing with H1 grease.
         i. Look for water seeping out of the bearing and continue greasing until the grease is visible.
         ii. Wipe the bearing inside and outside to prevent product contamination.

(___) 12. Check Product Zone for loose or peeling paint on motors, gearboxes, supports and wall and ceiings. (Product Zone is any area above where product is exposed.)
(___) 13. Check Product Zone for loose or missing caulk. Product Zone is any area above where product is exposed.)
(___) 14. Check Product Zone for rusty motors, gearboxes, supports, and walls and ceilings. (Product Zone is any area above where product is exposed.)

Figure 5-6 cont. Weekly Operating Meat Applicator, Final Changes

third attempt was made. It still lacked the necessary detail. So changes were made and sent back to the planner. Figure 5-6 highlights the final changes that needed to be made.

Even though this preventive maintenance procedure was greatly improved, there are still unanswered questions about the preventive maintenance details. Some of these questions are underlined in the preventive maintenance task detail. If these questions can be answered in the next iteration of the preventive maintenance task, then the model task should be finalized. This would provide the standard for which all other preventive maintenance tasks should be judged against.

Although this level of detail may seem to be overkill, it can be seen that the document can be used to train apprentice-level individuals to perform the preventive maintenance task properly. In addition, if they need training to perform the tasks, it will be easier to identify this need.

A Shutdown Preventive Maintenance Task

The first preventive maintenance task was an operating task. The next example, illustrated in Figure 5-7, is a task where the equipment must be shut down in order for it to be performed properly.

WEEKLY DOWN PM FOR THE MEAT APPLICATOR
INITIAL EACH ITEM AFTER COMPLETING:

(___)   1. Inspect the meat leveling rake, check gearbox, shafts, bearings,and mechanism. Check that rake tines are present.

(___)   2. Inspect nylon disc for wear. Check rake connecting rods for tightness and for signs of wear.

(___)   3. Inspect the beater bar rake, shaft, gearbox, coupling and bearings. Check that all tines are present

(___)   4. Check shafts, rollers, couplings, bearings and guides. Replace roundthane belts if needed.

(___)   5. Check Product Zone for loose or peeling paint on motors, gearboxes, supports, and walls and ceilings. Check Product Zone for loose or missing caulk. Check Product Zone for rusty motors, gearboxes, supports, and walls and ceilings. (Product Zone is any area above where product is exposed.)

Comments:

Figure 5-7 Weekly Down PM for the Meat Applicator

This first and rather obvious problem with this preventive maintenance task is the lack of any lock-out / tag-out instructions. In addition, as was the case with the previous preventive maintenance task, the details were severely lacking. So again, the planner was given instructions on how to modify the preventive maintenance task. Figure 5-8 shows the second round results.

WEEKLY DOWN PM FOR THE MEAT APPLICATOR
INITIAL EACH ITEM AFTER COMPLETING:

(__)   1.   Inspect the meat leveling rake, and check the gearbox for the following:
　　　　　Loose base bolts
　　　　　Leaking shaft seals
　　　　　Loose Housing bolts
　　　　　Proper lubricant level
　　　　　Physical damage
　　　　　???????
　　　　Check the shafts for the following:
　　　　　Alignment
　　　　　???????
　　　　Check the bearings for the following:
　　　　　Excessive play **(How much?)**
　　　　　Proper Lubrication?
　　　　　?????
　　　　Check the mechanism for the following:
　　　　　??????
　　　　Check that rake tines are present
　　　　　How many??
　　　　Inspect nylon disc for wear
　　　　　How to determine??
　　　　Check rake connecting rods for:
　　　　　Tightness **(Torque setting?)**
　　　　　Signs of wear **(Examples??)**

(__)   2.   Inspect the following items:
　　　　　Beater bar rake
　　　　　　For???
　　　　　Shaft
　　　　　　For???
　　　　　Gearbox
　　　　　　For???

continued on next page

Figure 5-8 Weekly Down PM for the Meat Applicator, 2nd Attempt

Coupling
For???
Bearings
For???
Check that all tines are present
How Many??

(____)   3.   Check the following:
Shafts
For????
Rollers
For????
Couplings
For????
Bearings
For????
Guides
For????
Replace roundthane belts if needed
How do we know if it needs changed????

(____)   4.   Check Product Zone for loose or peeling paint on motors, gearboxes, supports, and walls and ceilings.

(____)   5.   Check Product Zone for loose or missing caulk.

(____)   6.   Check Product Zone for rusty motors, gearboxes, supports, and walls and ceilings. (Product Zone is any area above where product is exposed.)

Additional Observations:

___

While the preventive maintenance task showed some improvement, there were still many details lacking. This were noted with questions (highlighted and underlined). The plan was sent back to the planner for further detail. Figure 5-9 shows the result.

Notice the improvements in the level of detail for the preventive maintenance procedure. Even though there is much more detail, there is still some opportunity for additional detail. However, consider how easy it would be to develop a training document from this level of detail. Also, if the goal was sometime in the future to transfer some of the maintenance tasks to the operators, this level of detail would make it easy to develop one point lessons or simple task sheets.

WEEKLY DOWN PM FOR THE MEAT APPLICATOR

Note this is a down preventive maintenance task. Lock out the following electrical switches at control panel 4-B.

Switch # 5, 7, 8.

This panel is beside the west side of the meat applicator. Make sure the green lights are illuminated before beginning this preventive maintenance task.

After locking out the switches, try the jog button on the meat applicators operator's panel to insure the equipment is de-energized.

INITIAL EACH ITEM AFTER COMPLETING:

(____) 1. Inspect the meat leveling rake.
    a. Check the gearbox for the following:
        i. Loose base bolts
        ii. Leaking shaft seals
        iii. Leaking housing seals
        iv. Loose housing bolts
        v. Proper lubricant level
        vi. Physical damage.
        vii. Insure that key is present.
        viii. Drain a small amount (one 8-oz test jar) of oil to check for water contamination.

    b. Check the shafts for the following:
        i. Alignment
        ii. Check the connecting arm bolts for tightness.
            25 ft-lbs.
        iii. Insure that each rake is in the vertical plane.
        iv. Check the bearings for the following:
            a) Excessive play—if there is visible movement in the races or between the inner race and shaft, write a PM Follow-up notification for replacement.

    c. **Proper Lubrication**
        i. Grease the bearing with H1 grease.
            a) Look for water seeping out of the bearing and continue greasing until the grease is visible.
        ii. Wipe the bearing inside and outside to prevent product contamination.

Figure 5-9 Weekly Down PM for the Meat Applicator, Further Detail

d. Check the setscrews for tightness.
    25 in-lbs)
e. Check the mechanism for the following:
    i. Inspect nylon disc for wear.
        a) If signs of wear are present (Frayed edges ?" thickness or less), check applicator side plates for at least 1/8" clearance.
        b) If disc is frayed or gouged, write a PM Follow-up notification to smooth out or replace depending on amount of damage (less than 1/4th inch of nylon on the edges)
f) Check rake connecting rods for:
    i. Tightness
        a) 25 ft-lbs
    ii.Check that all the rake tines are present.
        b) If 3 or more tines are broken or missing, write a PM Follow-up notification for replacement

(_____)   2.   Inspect the following items:
a. Beater bar rake
    i.Check that all tines are in rake.
    a) If 6 or more pins are missing, write a PM Follow-up notification for repair.
    ii. Check that rake is not bent. A bent rake will lope and could contact the belt, causing damage.
b. Shaft
    i. Inspect the fit between the shaft and inner race of support bearing.
c. Check the bearings for the following:
    i. Excessive play
    a) if there is visible movement in the races or between the inner race and shaft, write a PM Follow-up notification for replacement
    ii. Proper Lubrication
    a) Grease the bearing with H1 grease.
    b) Look for water seeping out of the bearing and continue greasing until the grease is visible.
    c) Wipe the bearing inside and outside to prevent product contamination.
    iii. Check the setscrews for tightness. (25 in-lbs)
d. Gearbox – Check for the following:
    i. Loose base bolts
    ii. Leaking shaft seals
    iii. Leaking housing seals
    iv. Loose housing bolts
    v. Proper lubricant level (1/2 full on sight glass)
    vi. Physical damage
    vii. Insure that key is present
    viii. Drain a small amount (one 8-oz test jar) of oil to check for water contamination.

  e. Coupling
    i. Check alignment of coupling with a straight edge.
    ii. Check coupling halves for damage or wear
    iii. Check flexible spider or boot for wear or damage
    iv. Check set screws for tightness. (25 in-lbs)
  f. Bearings
    i. Excessive play
      a) If there is visible movement in the races or between the inner race and shaft, write a PM Follow-up notification for replacement
    ii. Proper Lubrication
      a) Grease the bearing with H1 grease.
      b) Look for water seeping out of the bearing and continue greasing until the grease is visible.
      c) Wipe the bearing inside and outside to prevent product contamination.
    iii. Check the setscrews for tightness. (25 in-lbs)
  g. Check that all tines are present.
    Check that all the rake tines are present.
    If 3 or more tines are broken or missing,
    write a PM Follow-up notification

(_____) 3. Check the following:
  a. Shafts
    i. Inspect the fit between the shaft and inner race of support bearing.
    ii. Check key and key way of drive.
  b. Rollers
    i. Check that rollers are not bent.
    ii. Check that coated rollers do not have any missing coating
    iii. Check that V-guide at roller center is not obstructed.
  c. Couplings
    i. Check alignment of coupling with a straight edge.
    ii. Check coupling halves for damage or wear
    iii. Check flexible spider or boot for wear or damage
    iv. Check set screws for tightness.
  d. Bearings
    i. Excessive play
      a) if there is visible movement in the races or between the inner race and shaft, write a PM Follow-up notification for replacement
    ii. Proper Lubrication
      a) Grease the bearing with H1 grease.
      b) Look for water seeping out of the bearing and continue greasing until the grease is visible.

          iii.   Check the setscrews for tightness. (25 in-lbs)
    e. Guides
        i. Check that belt guides are not so tight that they wear the edge of belts.
        ii. V-guides on bottom side of belt should be tightly attached with no signs of coming loose or cracks.
        iii. If V-guide is loose less than 2 inches, trim loose section off with a sharp knife.
        iv. If loose section is greater than 2 inches, write a PM Follow-up notification
    f. Inspect and Replace roundthane belts if needed.
     i If roundthane belts are sagging or differences in tension is obvious, belts need be replaced.
        ii. Do not try to shorten old roundthanes, always replace with new material.
        iii. Cut roundthanes (???) inches and weld according to instructions included in welding kit.
        iv. Trim excess material from weld site.

(_____)   4. Check Product Zone for loose or peeling paint on motors, gearboxes, supports, and walls and ceilings.

(_____)   5. Check Product Zone for loose or missing caulk.

(_____)   6. Check Product Zone for rusty motors, gearboxes, supports, and walls and ceilings.

(Product Zone is any area above where product is exposed.)

When finished, unlock the following electrical switches at control panel 4-B. Switch # 5, 7, 8.

This panel is beside the west side of the meat applicator. Make sure the red lights are illuminated before notifying production the equipment is ready to operate.

After unlocking the switches, try the jog button on the meat applicators operator's panel to insure the equipment is energized and ready to operate.

Additional Observations:

    Although the exercise of developing this level of detail may seem laborious, it will provide major benefits as the organization matures its preventive maintenance program.

- **Estimates on PMs, should include:**
  - Preparation time
  - Travel time
  - OSHA policies
  - Security restrictions
  - Actual time to perform
  - Clean up time
  - Clerical time?

Figure 5-10 PM Estimates

Time Estimations

Once this level of detail has been developed, it is necessary to determine how long it will take to perform the preventive maintenance task. Figure 510 presents some of the considerations that need to be examined when determining the estimates. Of the items listed in Figure 5-10, the OSHA policies are especially important. Lock-out / tag-out instructions, confined space entry, or guarding instructions can have a dramatic impact on the time required to perform a preventive maintenance task. In some organizations, there are security requirements that consume a large amount of time due to the coordination necessary.

The clean up time and the clerical time to record the work that was performed or any identified new work is also critical if the preventive maintenance program is to show results. Without the details being properly recorded, the ability to improve the preventive maintenance program will be severely impacted. Accurate data will not be available.

The development of accurate estimates is dependent on all of the steps previously covered in this chapter. Without knowing what type of task it is going to be, the steps necessary to perform the work, accurate estimates will be difficult. An additional consideration for developing accurate estimates is the skill of the individual who will be performing the work. This will be the topic of the next chapter.

Determining the Skill Requirements for PM Tasks

Once the preventive maintenance tasks have been detailed, the next step is to determine the skills that will be required to perform the tasks. There are four main groups that will be responsible for performing the preventive maintenance tasks. They are listed in Figure 6-1.

Skill Requirements

- Operators
- Apprentices
- Journeymen
- Outside Contractors

Figure 6-1 Skill Groups

Operator Preventive Maintenance Tasks

Operator-based preventive maintenance tasks are generally related to the following activities:
- Taking operating readings
- Performing "sensory" inspections of the equipment
- Performing minor adjustments or service to the equipment

In most organizations, the operator-performed preventive maintenance tasks will make up anywhere from 10–40% of the overall preventive maintenance program. There are some companies where the operator and maintenance technicians have a shared role; in these cases, the percentage can be greater. However, for this model to be successful, the company will invest heavily in technical training, raising the technical skills of operators to at least an apprentice level. (This model will be discussed in detail in Volume 6 of the maintenance strategy series.)

Operators of the equipment are usually required to make operational checks of the equipment while it is running. These checks usually involve recording data over various process parameters. These will include temperature, flows, quality information, tolerances, etc. This process provides an opportunity to have them also record any information that will add value to the maintenance of the equipment, for example:

- Lubrication system flows and pressures
- Hydraulic system flows and pressures
- System temperatures

All of these parameters are important when determining if the equipment needs servicing or if a problem is developing. If the operator can be trained to record these readings properly, then the need to have a maintenance technician walk the same route as an operator is eliminated. This frees up maintenance resources to be re-deployed on higher level activities, such as predictive maintenance and reliability analysis.

When operators perform sensory inspections of the equipment, they are using their five senses to help determine the equipment's condition. This means the inspections should have them:

- Listening for specific sounds
  (Described on the inspection)
- Detecting unusual odors
- Touching the equipment for unusual vibration
- Visual inspection for leaks or other unusual conditions

With these inspections supplementing the maintenance services, the preventive maintenance program should be able to keep the equipment in an acceptable baseline condition.

If the operators are going to be involved in performing minor adjustments or servicing the equipment, they will require technical training. One of the major mistakes companies make when involving operators in this area is their they failure to provide the training to make the operator's efforts successful. For example, if operators are to lubricate the equipment, they should be trained to insure:

- They apply the correct lubricant

- They apply the correct amount
- They use the correct application method
- They know where each fitting is located
- They apply the lubricant at the correct frequency
- They clean the fitting before and after lubricating it

Unfortunately, most companies just hand the operators a grease gun and tell them to lubricate the equipment. This results in damage to the equipment and eventual downtime and higher maintenance costs.

If the operators are to be involved in preventive maintenance activities, proper documentation and training are essential. Otherwise, the lack of results will quickly end their involvement in the preventive maintenance program.

Craft Apprentice Preventive Maintenance Tasks

In addition to the operator-level preventive maintenance tasks, there are also those tasks that can be performed by apprentice-level craft technicians. The goal in developing this level of preventive maintenance task would be to free up journeyman-level craft resources to be able to focus on higher-level, more technical maintenance activities.

What type of tasks would be appropriate for apprentice-level craft technicians? It would be those tasks that are routine in nature, such as those not requiring advance technical skills to inspect. These tasks would likely exclude any that required extensive root cause analysis skills to perform a preventive maintenance inspection. Apprentice-level craft technicians should be able to perform acceptable / not acceptable inspections. However, the task required them to make a determination of what was causing the component being inspected to require repair or replacement, then it would be better to assign a journeyman-level technician to perform the task.

As noted previously, if the preventive maintenance tasks are properly detailed, the apprentice craft technicians can learn much about their equipment, allowing them to advance to a journeyman-level qualification at a more rapid rate. This, coupled with their formal training program, will allow the company to benefit as their skills are enhanced. The increased rate of progression will provide the company with a larger pool of skilled technicians.

Journeyman Level, Craft Technician Preventive Maintenance Tasks

In theory, journeymen should be able to perform any preventive maintenance task in their particular craft discipline. Their skills should allow them to properly service any equipment, determining its condition, and determining the needed service. They should be able to recognize any condition requiring additional work that is outside the scope of the preventive maintenance task and properly document the identified work for additional follow-up.

This is in theory. In actual practice, each journeyman craft technician has strengths and weaknesses. All should be properly skilled to perform preventive maintenance tasks to a level of detail that allows the task to be successful in properly servicing the equipment.

Before any preventive maintenance task is given this classification, it should be reviewed to insure it cannot be performed by operators or by apprentice-level maintenance technicians. Journeyman-level resources are at a premium in most organizations today and should be carefully deployed.

Outside Contractor Preventive Maintenance Tasks

In many companies, the use of outside contractors to perform preventive maintenance tasks has increased. This is particularly true in facilities-based organizations. These organizations will pick preventive maintenance tasks requiring specific skills, such as refrigeration or heating, ventilation, and air conditioning. For these organizations, the preventive maintenance labor load does not require full time resources to properly perform the tasks. In these cases, contracting out the tasks is financially prudent.

Determining the Proper Resources

How do companies make the correct decision on the resources to perform the preventive maintenance tasks? It should be an economic decision. The lower that the skill level is of the individual who can acceptably perform the tasks, the lower the overall costs will be. This is one of the major factors to consider when designing the preventive maintenance tasks. Although each organization will vary in the percentage distribution, there should be an initial goal set for the amount of the preventive maintenance program for which each group will be responsible. For example, a typical division might be:

- 20% by operators
- 50% by apprentice-level craft technicians
- 20% by journeyman-level craft technicians
- 10% by outside contract resources

The initial goal can be set for the preventive maintenance program before the actual development begins. This will help focus the development efforts towards a labor distribution goal. As the preventive maintenance development progresses, it will be necessary to re-examine the initial goals to insure they were realistic. If some adjustment, due to the technical level of the equipment or skills of the workforce, is required, the adjustment should be made.

Once the labor resources have been determined, the next step will be do determine the spare parts that will be required to perform the preventive maintenance task. This step is discussed in the following chapter.

7

DETERMINING THE PARTS REQUIREMENTS FOR PM TASKS

In this step, all of the spare parts are specified for the preventive maintenance tasks. This specification will allow the tasks to be performed in an efficient manner. Note that this step is only necessary if the preventive maintenance tasks are replacement oriented and not just inspection oriented. Inspection-oriented preventive maintenance tasks will not require spare parts to be detailed on the check sheet.

Bill of Materials

The bill of materials for each equipment item is typically developed during the implementation of the company's CMMS (Computerized Maintenance Management System) or its EAM (Enterprise Asset Management) System. The bill of materials (BOM) provides a way of looking into the plant spare part inventory to see what spare parts are being stocked for a particular piece of equipment.

The problem that most CMMS or EAM systems have with using the bill of material feature is that companies never load the BOM when implementing their systems. This is due to the fact that once the entire inventory is in their CMMS or EAM system, they must then take the time to go through the inventory database and electronically attach each part to each piece of equipment where it is used. This time-consuming task is typically omitted during a CMMS or EAM system implementation because it is viewed as being too resource intensive. However, this feature can save the maintenance organization a considerable amount of resources over time.

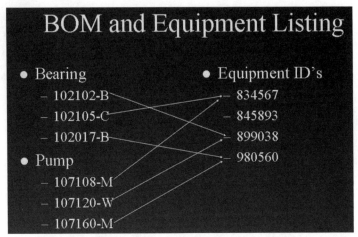

Figure 7-1 BOM and Equipment Listing

The BOM lists by equipment item all of the spare parts that are carried in stock in the company stores. There is a corresponding "where used" function that looks at the relationship in reverse, that is with the spare parts being listed and then all of the equipment where the spare part is utilized. In most CMMS or EAM systems, this approach is just a different way of viewing the same database. This relationship is shown in Figure 7-1. As can be seen, the spare parts are attached electronically to the equipment item on which they are used.

If a list of spare parts was produced for a particular piece of equipment, it would appear similar to the sample in Figure 7-2.

The list produces a description of the spare part, the manufacturer's part number (if known or available), and the company part number. Thus, when any individuals developing the preventive maintenance task bring up the equipment BOM, they can select and add any part on the list to the preventive maintenance task part listing.

If the list like the one shown in Figure 7-2 was used to develop the spare parts list for a preventive maintenance task, the planners could easily work down the list, selecting any bearings or belts that might be required for the equipment to be successfully serviced.

Once finished and attached to the preventive maintenance task, the list would give a print out as seen in Figure 7-3.

When the craft technicians receive the preventive maintenance task sheet, the parts are listed on the sheet. It is then a simple matter for the technicians to walk to the storeroom, order the parts, and carry out the preventive maintenance task.

Par	OEM#	Internal #
ANVIL		103567
ASSY, FIN WHEEL		117649
BASE, FORMER		151222
BEARING, INFEED AND OUTFEED ROLLERS		10008
BEARING, FIN WHEEL		116849
BEARING, FIN WHEEL		100434
BEARING, FILM ROLLER ASSEMBLY	503442	151045
BEARING, FILM ROLLER ASSEMBLY	182129	151046
BEARING, FLANGE	463996	SEE 150802
BEARING, SLEEVE ROLLER	521-49323-89	151040
BEARING, THRUST OIL TRI-TRONICS		151041
BELT, DRIVE AUXILARY FIN WHEEL		107902
BELT, DRIVE CRIMPER INSIDE PANEL NON OP SIDE		111303
BELT, CONVEYOR SPLIT		150565
**BACK-UP SPLIT BELT; MODIFY WIDTH		131605
BELT, DRIVE INSIDE CRIMPER		146099
BOLT, FIN WHEEL-CUT ?-13		146099
BRACE, FORMER SUPPOR	08212234	151130
CONTROLLER, MOTION PMAC		131382
CONTROLLER, WATLOW UD1A-1CES-0000		131291
COLLECTOR, 2-RING MOUNT FOR		
LARGE OPENING COMUTATOR		117649
COLLECTOR, 4-RING MOUNT FOR		
LARGE OPENING COMUTATOR		103269
CRIMPER, UPPER		145045
CRIMPER, LOWER		145044
DISH, FIN WHEEL	291311	118700
DRIVE, SERVO	440331	131281
ELECTRO-SAFE, FMO	106257	127242
EYE, OPTIC FILM REGISTRATION FIBER		146022
EYE, SMART EDR-CMS-2BV1G		150466 G-5-2
ROLLER, FILM PINCH		161415
FILM POWER FEED-dr20-bkst	384725	151042
FILTER, AIR CONDITIONER		131123
HANDLE, ACCESS DOOR		267810
HEATER, CRIMPER		145043 K-4-2
HEAD, CRIMPER GEARBOX/ALPHA		145758
HINGE, GUARD DOOR CRIMPER HEAD		146169
HOLDER, BRUSH 2 RING ASSY		131207
HOLDER, BRUSH 4 RING ASSY		131208

continued on next page

Figure 7-2   Sample Spares Part List

Par	OEM#	Internal #
LATCH-GAP CONVEYOR SIDE COVERS, SMOOTH WELD		151017
LOC-K-4-3		145794
PANEL, CRIMPER NON-OP SIDE		111303
PLATE, BASE FORMER		151222
PLATE, LEFT SIDE LOC	K-4-7	151133
PLATE, RIGHT SIDE LOC	K-4-6	151132
PLATE, WING FORMER TOP MAIN		151131
PULLEY, DRIVE TRANS	464889	150803
PULLEY, DRIVE TAPER LOC 1210 0.75		151044
PULLEY, TIMING 40L050	181064	151043
RING, COLLECTOR 4 RING	305851	111638
RING, COLLECTOR 2 RING	305856	
ROLLER, BELT INFEED AND OUTFEED		175140
ROLLER, PINCH FOR FILM		161415
ROLLER, TRACKING RUBERIZED USDA		173786
SAFETY, NON-OP SIDEDOOR		111303
SAFETY, DOOR		130751
SHAFT, FIN WHEEL INSIDE	324592	100435
SHOE, BRAKE ASSY	222152	150021
SPACER, FIN WHEEL	320048	
SPRING, FIN WHEEL COMMENTATOR	082-39032	118596
SPROCKET, REVERSE TAPER 40b 14RM-1008		173967
VALVE, REJECT SOL MAC		128583 G-3-8
VALVE, FIN WHEEL	McMaster #4891k71	
WASHER, FIN WHEEL		173571
WING, FORMER		151233
WING, FORMER		151224
WIRES, FIN WHEEL COMMENTATOR	082-40465	101667

Figure 7-2   Sample Spares Part List continued

BEARING, FILM ROLLER ASSEMBLY	182129	151046
BEARING, FIN WHEEL		116849
ROLLER, BELT INFEED AND OUTFEED		175140
SPROCKET, REVERSE TAPER 40b 14RM-1008		173967

Figure 7-3 Sample Specific Parts List

In more advanced environments, once the preventive maintenance task is scheduled, the storeroom is notified electronically and the spare parts are kitted and staged. When the craft technicians walk to the storeroom, they simply give the storeroom attendant the work order number for the preventive maintenance task. The attendant is able to give the technicians the staged parts almost instantly. This process further reduces the delay time in obtaining parts needed to perform the preventive maintenance task. The definition of kitting is highlighted in Figure 7-4.

- **The process of obtaining all materials for a specific job before it is started.**
- **This requires effective planning of the maintenance tasks, whether:**
  - Preventive maintenance or
  - Routine maintenance

Figure 7-4 Kitting and Staging

One additional step can be taken to improve the efficiency of obtaining parts and increasing the efficiency of the craft technicians. That step is to deliver the spare parts for the preventive maintenance task to the job site the shift before the work is to be performed. This action is typically done on geographically-dispersed sites where the travel time would be too extensive for the technicians to go to the storeroom for their spare parts for each preventive maintenance task to be performed.

If these processes are followed when using spare parts to perform preventive maintenance tasks, the efficiency of the work performed will be increased. Greater efficiency will allow more preventive tasks to be completed using the same resources.

8

DETERMINING THE SCHEDULING REQUIREMENTS FOR PM TASKS

Several considerations must be reviewed when determining the schedule for a preventive maintenance program. These considerations are listed in Figure 8-1.

- **Regulatory**
  - Non-Negotiable
- **Fixed versus Dynamic (or Sliding)**
- **Calendar/ Usage/ Condition Based**
- **Inspections or Task Based**

Figure 8-1 Scheduling Preventive Maintenance

Scheduling is a two-step process. It includes not only how long each task should take, but how often it should be performed. In determining how long it takes to perform a task, the estimate should consider:

- Time required getting tools and materials ready for the job
- Travel time to get to the job
- Any safety, environmental, or hazardous materials restrictions
- How long it actually takes to perform the task
- How long it takes to clean up the area and put all tools and materials away

If good estimating techniques are used, then the scheduling and completion of the tasks is much more accurate. However, one factor can skew the schedules for PM programs—the time someone spends on the job performing work that is not on the PM inspection or service. As craft technicians service or inspect an equipment unit, they will occasionally find a problem that is beginning to develop. The question is how long should they spend correcting the problem before they ask for help, or write a work order to correct the problem. This is generally a policy decision that should be made when starting a PM program.

There are two factors to consider when making the decision. The first is the time it would take to come back to the dispatching point and write a work order to have the work done. Depending of the geography of the plant, the travel time could be considerable and should be taken into consideration. The other extreme is the damage a prolonged task would cause to the PM schedule. If a task is estimated for four hours, but takes eight hours to perform because of the other problems encountered, the schedule will suffer. Because the time is charged to a PM work order, performing what should be routine repairs and charging them to the PM charge number will inflate the PM costs, hiding the true repair costs.

In most cases, companies begin with a time limit of an hour. Any additional work could be performed up to the limit of an hour. If the work requires more than an hour, then the craft technician should come back and write a work order to perform the work. This guideline allows for planning and scheduling of the work, making the work more effective.

This point leads into a second: the scheduling of the PM program. Scheduling PM tasks depends on what type of PM is specified. For example, OSHA inspections, environmental equipment, and hazardous equip-

---

Categorize PM tasks and determine schedule

– **Mandatory PMs** are performed at all costs when they come due
  • OSHA-compliance inspections
  • Inspection and servicing of environmental and hazardous equipment
  • Safety inspections, EPA inspections, License inspections
– If these PMs are missed the company is liable for fines
– Mandatory PMs are usually fixed-frequency and *cannot be altered or ignored*
– **Non-mandatory PMs** can be postponed for a short time period without resulting in immediate failure or penalty
– Inspections or Service?

Figure 8-2 Categorize PM tasks and determine schedule

ment all require certain servicing on specified and regulated intervals. Figure 8-2 highlights this. If these PMs are missed for any reason, and the regulatory government agency checks, the company would be liable for fines. These tasks are classified as mandatory PMs; they must be performed or something damaging to the company, equipment, or personnel will happen. They are usually fixed frequency PMs and cannot be altered. A non-mandatory PM is one that does not involve regulatory considerations, even though by missing it, the equipment will experience performance problems.

Another way of classifying PMs is by how they are scheduled. A typical classification is pyramiding or non-pyramiding. A pyramiding PM (see Figure 8-3) is one that is due but is not completed in the allotted time window. As a fixed frequency PM, when it comes due a second time, another work order is issued. Thus, the work order pyramids. Non-pyramiding work orders are not issued a second time. They do not issue until the first work order is complete. With fixed frequency PMs, the next due date slides based on the completion date, not the true next due date.

Categorize PM tasks and determine schedule

– **Pyramiding PMs** are non-mandatory, fixed-interval PMs that are generated each time they come due
  • When a PM is late, and the next one for the same work comes due, the first one should be canceled with a comment the PM was skipped
    – The new PM should carry the due date from the canceled PM, so that it is understood how overdue the task is

– **Floating or non-pyramiding PMs** are based on last completion date
  • Uncompleted PM is thrown away and the new one is scheduled.
  • No notification that any PMs were missed

Figure 8-3 Pyramiding and Floating PMs

With the pyramiding PM due on a fixed frequency, it is necessary to write a cancellation or missed notice on any uncompleted PMs. In this example, five completions would be noted, with two more being noted as missed. With the non-pyramiding PMs, there are also five completions during the same time period. However, when a failure occurs and the PM program is checked, the non-pyramiding PMs will show no missed tasks. This incomplete information leads the maintenance department to look somewhere else for the solution to the problem, when the real fault is with the PM program. Non-pyramiding PMs can hide potential equipment problems.

Another decision point in the program is whether to make the PMs just inspections or task oriented (see Figure 8-4). If they are just inspections, the inspectors must come back to the dispatch point and write the work orders for someone else to go out and perform the work. This can lead to a rift between the inspectors and the rest of the maintenance workforce. The task-oriented PMs instruct the inspectors to make minor repairs and allow time for them to do this. When making this decision, the future must be kept in mind. If complex tasks are specified, it makes turning the PMs over to the operators more difficult. If the future direction is TPM, then the tasks must be set up and designed with that goal in mind. Otherwise the transfer of any of the maintenance tasks to operations will necessitate a complete re-write of the TPM program.

Categorize PM tasks and determine schedule

– **Inspections** will only involve filling out a check sheet and then writing work orders to cover any problems discovered during the inspection.

– **Task-oriented** PMs allow the individual performing the PM to take time to make minor repairs or adjustments

  • Establish time limit set on how long each task should take.

  • This prevents the PM program from accumulating labor costs that should be attributed to routine repairs.

Figure 8-4 Inspections

Advanced Preventive Maintenance Scheduling

An advanced PM scheduling technique is the total cost strategy. This scheduling strategy involves the financial impact a preventive maintenance task has on the operation of the equipment. It is necessary to put the costs vs. benefit discussion in a form where all parties involved can understand it. Figure 8-5 highlights a graphic format of the total cost strategy. The figure shows that the decision for scheduling a preventive maintenance task would be made, not on what is best for the operations group, nor on what is best for the maintenance group, but what is the lowest combined cost. This is the effective "bottom line" for the company. This is the type of decision that companies must make if they are to optimize their resources.

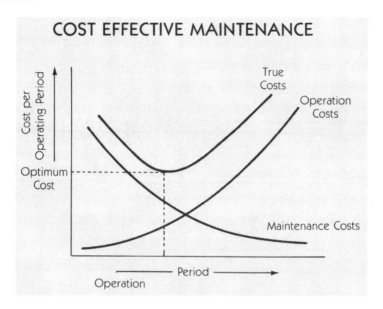

COST EFFECTIVE MAINTENANCE

Figure 8-5

How does a company go about collecting the information required to perform this type of preventive maintenance scheduling? It starts with assigning a cost to downtime. It may be useful to use the financial or accounting departments to find out what an hour or a shift of lost production is worth for a piece of equipment. This cost might include lost sales, employee salaries and overhead, the cost to make up lost production (if it can be made up), and any measurable depreciation to the assets. The figures coming from the financial department will usually be conservative, but will not be disputed by other parts of the organization.

With these figures agreed to, it is necessary to understand the maintenance costs involved. These costs may include the labor, material or supply, and miscellaneous costs that will be incurred due to the repair or the failure. Both costs may be needed to compare an overhaul to a run-to-failure approach to maintenance. Additional costs that may be incurred should also be calculated. These may include the hazardous materials, EPA, OSHA, or safety considerations.

When examining preventive maintenance frequencies, an example

BASIC MAINTENANCE COST CALCULATION

Repair Cost = $1,500

If the pump was serviced once every 100 hours, the cost would be:

$$\frac{1500}{100} = \$15.00/hr$$

If the pump was serviced once every 500 hours, the cost would be:

$$\frac{1500}{500} = \$3.00/hr$$

The table version would be:

Service Frequency (Hours)	Maintenance Cost (Dollars)
100	15.00
500	3.00
1000	1.50
1500	1.00
2000	0.75
2500	0.60
3000	0.50
3500	0.43
4000	0.38

Figure 8-6

common to most plants or facilities is a centrifugal pump. This pump may be pumping a product or moving cooling water. The point is it will have a value to its service. Setting a price on the value gives a reference from which to start. If the value is $100.00 per hour, then this amount forms the basis for the following preventive maintenance scheduling problem.

For this example, it costs $1,500.00 dollars for parts and labor to overhaul the pump. There is no downtime cost since there is a standby pump available. The pump performance is measured, and it is found that after 4000 hours of operation, it loses 5% of its capacity. It is assumed that the drop is linear and continues to be so throughout the life of the pump. The question is asked, when is it cost effective to perform a preventive maintenance task to remove the pump from service and clean the rotor?

LOST PERFORMANCE CALCULATION

If the performance loss is linear, then at 4000 hours of operation, the loss is 5% and the value is $100.00, then the value of the loss is:

$$0.05 \times \$100.00/hr = \$5.00/hr$$

Therefore, at 4000 hours of operation, the pump is producing only a value of $95.00, or it is losing $5.00 per hour.

The table version would be:

Time Since Last Service (Hours)	Lost Performance Cost (Dollars)
100	0.13
500	0.63
1000	1.25
1500	1.88
2000	2.50
2500	3.12
3000	3.74
3500	4.36
4000	5.00

Figure 8-7

The problem is solved by calculating the amount of maintenance cost versus lost performance cost per hour. The two are combined to give the lowest total cost. The techniques to perform this analysis are illustrated in Figure 8-6, 8-7, 8-8, and 8-9.

In Figure 8-6 the maintenance cost is calculated. The mistake of only considering maintenance costs is highlighted in this particular example. If only the maintenance costs were considered, it would be advisable to delay performing preventive maintenance on the pump for as long as possible.

The point in Figure 8-7, is that if you delay the preventive maintenance, the amount of lost production cost is increasing linearly. However, the difference between Figures 8-7 and 8-8 is that the latter takes into consideration the understanding that the performance fall off is triangular and not the total area volume of the rectangle.

The summary of the problem is in Figure 8-9. It plots the falling maintenance cost against the increasing cost of the lost performance.

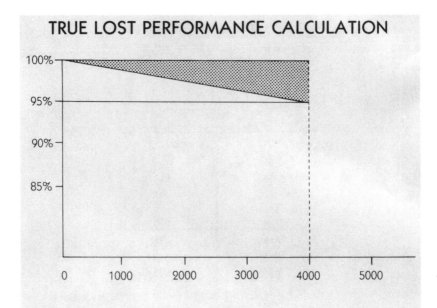

TRUE LOST PERFORMANCE CALCULATION

The total loss is not the entire rectangle, but only ½ of its area. So the loss would only be ½ of the calculated amount. The revised version of the table would be:

Time Since Last Service (Hours)	Lost Performance Cost (Dollars)
100	0.065
500	0.31
1000	0.63
1500	0.94
2000	1.25
2500	1.56
3000	1.87
3500	2.18
4000	2.50

Figure 8-8

These two factors added together will give the total of the true costs. The decision can then be made on lowest true cost, which in this problem would indicate that preventive maintenance action should be performed every 1500 hours of actual run time.

TRUE TOTAL COST CALCULATION

The true total cost can now be established for this task. Combining the tables results in:

Time Since Last Service (Hours)	Maintenance Cost (Dollars)	Lost Performance Cost (Dollars)	True Total Costs (Dollars)
100	15.00	0.065	15.065
500	3.00	0.31	3.31
1000	1.50	0.63	2.13
1500	1.00	0.94	1.94
2000	0.75	1.25	2.00
2500	0.60	1.56	2.16
3000	0.50	1.87	2.37
3500	0.43	2.18	2.61
4000	0.38	2.50	2.88

Figure 8-9

The problem can become more complex; for example, this problem has no penalty cost for the downtime required for the preventive maintenance action. If it would cost additional monies for the downtime, then this column would have to be added. However, in the example just given, the only downtime that would likely be incurred is when the repair was made.

Some problems will include breakdowns when maintenance intervals exceed a certain level. This means that downtime may have to be factored in during the cycle. This will radically alter the results of the calculation. For example, continuing to expand the problem in Figure 8-9, the downtime costs could be added, with the result being Figure 8-10. Now the downtime drives the true cost even higher than before. However, in this example, the initial cost of the downtime is factored in immediately.

The breakdown will only occur if the preventive maintenance is not performed before 3000 hours of operation. If the preventive maintenance frequency extends beyond that time, an additional cost of $2,400.00 (24 hours X $100.00 per hour value of the process) will be incurred. This removes any doubt that the lowest total cost would be around 2000 hours of operation, but definitely before 3000 hours of operation.

The ability to apply statistical techniques can go far beyond the simple example used in the text. Consider the ability to perform this type

FACTORING IN BREAKDOWN COSTS

If a breakdown occurs and there is no spare, then the service also results in a cost penalty to the Operations Group.

— For planned service—8 hours of downtime
— If a breakdown occurs—24 hours of downtime

A breakdown will occur every 3000 hours of operation, based on repair records.

The table shows:

Time Since Last Service (Hours)	Maintenance Cost (Dollars)	Lost Performance Cost (Dollars)	Downtime Cost (Dollars)	True Total Costs (Dollars)
100	15.00	0.065	8.00	23.07
500	3.00	0.31	1.60	4.91
1000	1.50	0.63	0.80	2.93
1500	1.00	0.94	0.53	2.47
2000	0.75	1.25	0.40	2.40
2500	0.60	1.56	0.32	2.48
3000	0.50	1.87	1.07	3.44
3500	0.43	2.18	0.91	3.52
4000	0.38	2.50	0.80	3.68

Figure 8-10

of costing for each sub-component of a large mechanical drive; the ability to determine the lowest life cycle costing for complex equipment; or the ability to determine the amount of resources to be spent on redesign and retrofit engineering projects, based on anticipated return on the investment.

Given that this model is mathematically straightforward, why are there so few companies doing this type of calculations? It is because very few of them have any reliable data with which to work. The data to do these calculations must come from the work order system. Without this reliable data, companies will be back to estimating when preventive maintenance activities should take place and not financially optimizing their resources.

The scheduling of the preventive maintenance tasks can utilize strategies ranging from very simple calendar-based techniques to the very accurate, but more complex total cost methods. As the preventive mainte-nance programs mature, they will progressively become more focused on the financial impact of the program. Thus the total cost techniques will become part of the preventive maintenance scheduling strategy.

EXECUTING PREVENTIVE MAINTENANCE TASKS

Once the tasks have been properly developed and the schedule determined, the next step is the actual performance of the tasks. In Chapter 5, the topic of developing detailed preventive maintenance activities was covered. If the preventive maintenance tasks have been developed using these guidelines, the work execution should be a matter of the technicians following the directions on the check sheets.

If the guidelines have been followed, the maintenance technicians will be provided detailed instructions for the task being performed, the spare parts that are required, and any special tools, regulatory instructions, etc. It will simply be a matter of executing the assigned tasks.

If the task is assigned to an apprentice technician or to an operator, it may be necessary to observe them performing the preventive maintenance activity to insure that it is properly executed. Once their ability to perform the preventive maintenance task is validated, they should be able to complete it independent of an observer.

Technician Assignments

Although the preventive maintenance task has the job classification already assigned, it is up to the immediate supervisor to determine which employee will actually perform the work activity. What factors are used to determine which employee will be assigned to perform the preventive maintenance task? The first is the complexity of the task. Is this task something that is detailed enough that any employee can perform it independent of any supervision? If so, then it is only a matter of the supervisor assigning the individual employee.

What if the preventive maintenance task is a training opportunity for a specific employee? Then, depending on the supervisor's schedule, they may assign the preventive maintenance task to the technicians or operators needing the training and spend time coaching them through the task. Once the supervisor is satisfied that the individuals have demonstrated the skills necessary to perform the preventive maintenance task independently, this achievement should be noted in their training record.

What if the preventive maintenance task requires special skills in order to be performed correctly? Then the supervisor should inventory the crew to see if the required skills are scheduled. If the skills are not on the scheduled crew, the supervisor should contact the planner/scheduler and have the preventive maintenance task rescheduled when the technician with the proper skills will be available. It is essential that technicians with the appropriate skills perform the preventive maintenance tasks; otherwise the integrity of the program will be compromised.

Supervisory Activities

The activities required form the supervisor during the execution phase of the preventive maintenance program included training, coaching, and certification. If the employees are not skilled at the task, perhaps never having performed it before, the supervisor will need to be with them, walking them through the task, making sure the safety instructions, the work process, and goal for each step are clearly understood and performed.

If the employees have performed the preventive maintenance task before and simply need a few reminders, then coaching is probably more in order. A coaching session may also be advisable just prior to an employee being certified on a particular task.

When the supervisors are involved in the certification of employees to perform a preventive maintenance task, they will be required to watch the employee perform the task and insure they are capable of performing the task safely. The supervisors will also make sure the task is performed correctly from a technical aspect. The supervisors will then be required to sign off on the employees, validating they are qualified to perform the task without observation in the future.

The final activity that supervisors may be involved with is periodic audit of the preventive maintenance tasks to insure the employees are continuing to perform the task correctly, both from a safety and technical perspective. This activity insures the employees are not overlooking any of the preventive maintenance procedure that could compromise their safety or lead to damage of the equipment.

Although proper execution of the preventive maintenance task lies primarily with the operators or the maintenance technicians, supervisors have ultimate responsibility for insuring the integrity of the program.

If the suggestions in this chapter are followed, the preventive maintenance program should not experience any problems with the execution of the planned tasks.

10

CONSISTENT PM PROGRAM
FOLLOW-UP

Once the preventive maintenance program is properly developed and implemented, and the tasks are properly executed, how does the program continue to maintain focus and deliver the desired results? It begins by reviewing the focus of a preventive maintenance program.

The first step is realizing what types of failures can and cannot be reduced or eliminated with a preventive maintenance program. The first two types of failures that an organization will try but fail to improve with their preventive maintenance program are listed in Figure 10-1. They are:

- Infant mortality
- Random failures

PM Focus

- What types of equipment failure is it best to address with a PM program?
  - Infant mortality
    - Occurs in the first few hours of a component's life
    - Usually electronics
    - Impossible to PM
  - Random failures
    - Occur without notice or warning
    - Difficult to predict
    - Stem from an engineering- or materials-related flaw
    - Unpredictable, therefore, PM will not prevent them

Figure 10-1 Infant Mortality and Random Failures

Infant mortality failure is a failure that occurs almost immediately after the equipment is started. Electronic failures is one of the primary examples. The first time a load is applied on a component, it fails. There is no preventive maintenance that can be performed to prevent this type of failure. There are mechanical examples of this type of failure also, for example, when a pump or motor is started for the first time and a bearing fails. This failure can be from a manufacturing defect in the bearing, damage to the bearing during a rebuild, or even damage to the bearing in transit.

Infant mortality failures need to be properly identified so that preventive maintenance resources are not expended trying to prevent this failure. Although root cause analysis can be performed and the real reason for failure identified and eliminated, this would be more the responsibility of the maintenance engineer and outside the scope of the preventive maintenance program.

Eliminating random failures is also outside the scope of the preventive maintenance program. This type of failure occurs after an equipment component has been in operation for a time period and a component just fails. It is a failure with no perceived pattern or sign of impending failure. If the component was to be with other similar components historically, there would be no definite life cycle where they could be replaced proactively with any certainty of eliminating the failure and still be cost effective.

Figure 10-2 introduce the third type of failure that can not be eliminated by preventive maintenance, which is a failure due to abuse or misuse of the equipment. This type of failure can occur when equipment is

PM Focus

- What types of equipment failure is it best to address with a PM program?
  - **Abuse or misuse failures**
    - Training or attitude problem
    - No PM program can prevent this type of failure.
  - **Normal wear-out**
    - PM designed to address failures develop
    - Progressively over a relatively long period of time
    - PM will spot signs of wear
    - Measures to correct the situation scheduled.

Figure 10-2 Abuse or Misuse Failures and Normal Wear Out

operated beyond design speed or load. When the equipment is operated outside its design parameters, it is usually a management issue. This occurs when operations is trying to make up lost product or meet an overly-optimistic production schedule produced by someone within the organization who has not factored the equipment design capacity into the schedule. This area must be addressed by management guidelines and not by the preventive maintenance program.

In some cases, the operators will abuse or misuse the equipment. This problem is generally due to a training issue, which means the operators have not been properly trained on how to operate the equipment. Correcting this problem is a matter of obtaining the Standard Operating Procedures (SOPs), making sure the operators understand how the equipment should be operated, and observing the operators apply their new knowledge. Infrequently, operators will deliberately damage their equipment. If this is the case, then it becomes a Human Resources (HR) issue, not an issue for the preventive maintenance program.

Figure 10-2 also highlights the real focus of a preventive maintenance program: eliminating or preventing normal wear out failures. These failures are the types of failures that occur when a component is reaching the end of its design life. The key to successful preventive maintenance is to know when this time is and replace the component just before it fails. This is the most cost-effective form of preventive maintenance. The other key is to know when to service the component so that it

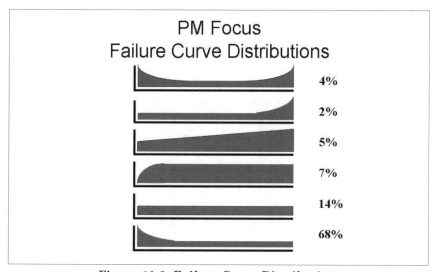

Figure 10-3 Failure Curve Distributions

· reaches its maximum design life. Finally, the preventive maintenance program should incorporate inspections, timed to insure that when wear is detected, the proper service can be provided to insure the component reaches its design life.

Figure 10-3 provides another looks at the failure curves that drive a preventive maintenance program. There is a considerable amount of misunderstanding about this set of failure curves in the maintenance and reliability disciplines. The curves pictured in Figure 10-3 detail the most common failure patterns for equipment *components*. The only curve that applies to equipment systems is the first curve, commonly referred to as the bathtub curve.

Many individuals will look at the last two failure curves in Figure 10-3 and conclude that because 82% failures occur without notice, there is little use in having a preventive maintenance program. Nothing could be further from the truth. This 82% of failures is at a *component* level and typically is applied to electronic devices.

However, these curves fail to consider the majority of mechanical and fluid power components that make up the majority of equipment *systems* that the electronic devices are actually controlling. In reality, these mechanical and fluid power systems follow the failure pattern set by the bathtub curve. In fact, studies have shown that the percentages are almost reversed when looking at the mechanical and fluid power systems. These numbers indicate that the vast majority of failures have some form of wear that can be trended or detected, allowing for the preventive maintenance service to increase the equipment life or detect the failure before it occurs, in turn, allowing for a proactive replacement of the components.

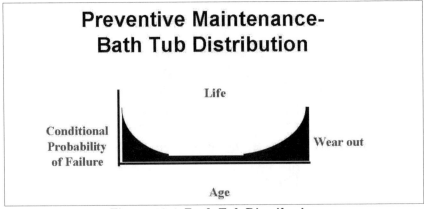

Figure 10-4 Bath Tub Distribution

Figure 10-4 shows the bathtub failure curve in greater detail. The first part of the curve is the infant mortality part of the curve usually encountered during the start-up or commissioning phase of the equipment. This is where the individual *components* may fail during burn-in. The overall equipment system shows a higher than normal failure rate, decreasing to the normal failure rate during this time.

Once the equipment system is into the normal part of the bathtub curve, the various components will fail at a combined steady rate, depicted by the flat portion of the bathtub curve. As the system components begin to reach the end of their life design life, the overall failure rate for the equipment system begins to show an increasing rate of failure. The equipment system has thus moved into the wear out phase of the bathtub curve.

It is at this point that decisions are made as to 1) the timing of an equipment overhaul to replace the components that are reaching the end of their design life and 2) resting the equipment system to an earlier point on the normal life (flat portion) of the bathtub curve. These adjustments allow for maximum life of the equipment system to be achieved. Figure 10-5 summarizes this scenario.

Preventive Maintenance Audits

It is advisable to review the preventive maintenance program annual for areas where it can be improved. The audit will involve examining

---

Bath Tub Curve

- As equipment becomes older, it requires closer attention to maintenance.
- Major overhauls or rebuilding may partially establish the curve.
- The majority of components do not follow the bath tub curve failure pattern; but equipment systems do...

Figure 10-5 Bath Tub Curve

the following details:

- The current preventive maintenance tasks
- The current schedule for the tasks
- The last year's failure history for the equipment
- The problem-cause-action results from the equipment history

Using this information (at a minimum), the preventive maintenance program should be reviewed to insure that there is not unnecessary preventive maintenance being performed on the equipment. For example, if a preventive maintenance task is being performed monthly, but the equipment is experiencing no problems, is it possible that the task could be performed on a quarterly basis? This review is particularly applicable to a preventive maintenance inspection.

If the preventive maintenance review shows that, despite the services being performed, the equipment is still breaking down or requiring reactive work, then perhaps the frequency of the work needs to be increased.

This level of preventive maintenance review can help to adjust the program so that it is cost effective and properly maintains the equipment. However, there are additional techniques that can be applied. These include a predictive analysis and a reliability analysis. (These techniques are covered in extensive detail in Volumes 7 and 8 of the maintenance strategy series.) A predictive analysis involves using a predictive tool (vibration analysis, sonics, thermography, or oil analysis) to insure that the components in a system are wearing at an acceptable rate. If rapid wear is detected, then the preventive maintenance program should be adjusted to compensate for the root cause of the wear.

A reliability analysis utilizes the equipment data to review the equipment performance in much greater detail. Engineering data, such as Mean Time Between Failures (MTBF) and Mean Time To Repair (MTTR) are used to evaluate the equipment performance. In part, this compares the preventive maintenance program to the maintenance needs of the equipment. This analysis will uncover where too much or too little preventive maintenance is being performed. By adjusting the preventive maintenance program, increased equipment life can be achieved.

Preventive maintenance is critical to operating a cost-effective maintenance department. A quote from another publication (The Complete Handbook of Maintenance Management by John Hientzelman,

PM Versus Breakdowns

- Repair and rehabilitation costs for damage to equipment under a breakdown program can be as high as 300% or more over preventive maintenance costs.
  − The Complete Handbook of Maintenance Management − John Heintzelman

Figure 10-6 PM versus Breakdowns

1976) makes the point shown in Figure 10-6. There are no shortcuts—the preventive maintenance program does form the foundation for all of the subsequent maintenance strategies. It can not treated in cursory manner if the rest of the maintenance strategies are going to be effective and efficient.

Performance Management for PM Programs

How can you monitor the effectiveness of the PM program? How can you know when the program needs adjustment to insure effectiveness? Observations can be made in several areas:

1. Low overall equipment effectiveness (using the formula)
2. Longer MTTR (mean time to repair)
3. Maintenance-related quality problems
4. Cost per repair increases
5. Rapid decrease in the value of capital assets

Low equipment effectiveness should be examined formula by formula. This could be particularly PM related when the availability is the lowering factor. If quality is the problem, then the quality part of the formula will highlight this point.

A longer MTTR (mean time to repair) indicates that a failure or breakdown takes a longer time to repair, meaning a more severe problem has been encountered. It takes longer to repair than a less serious problem that should have been found in its early stages by an effective PM program.

Comparing quality problems to the equipment effectiveness formula can help to spot those quality problems that are maintenance related. If the problems are related due to routine services or PMs, then the program's effectiveness is questionable. This indicator will allow corrective action to be taken.

The cost per repair is an indicator that shows repairs are more involved and taking more parts and labor than they should because of problems caused by a failure or an advanced stage of deterioration.

The rapid deterioration of assets simply means that the equipment and facilities are not lasting as long as their design intends. Lack of maintenance contributes to higher-than-normal capital expenditures for equipment replacements. If equipment is properly maintained, the effectiveness should be high enough to avoid purchasing replacements of backup equipment.

In developing any maintenance program, particularly TPM, it is essential to have a very effective PM program. This program eliminates the emergency or fire fighting maintenance that is costly in labor and materials. It also disrupts the relationship between operations and maintenance. It is imperative that the program be as effective as possible. If the PM program is ineffective, the rest of the program will suffer and eventually be discontinued. There have been many companies that have tried to replace their ineffective PM program with a TPM program. However, if they can't make a PM program effective, they will never make the TPM program successful.

Preventive Maintenance Program Key Performance Indicators

The first indicator highlights the impact the preventive maintenance program has on the plant or facility. It focuses on what the preventive maintenance program is designed to eliminate—equipment breakdowns.

1. Equipment Downtime Caused by Breakdowns

Downtime Caused By Breakdowns
Total Downtime   Expressed as a %

This indicator takes the total breakdown downtime for a piece of equipment, a department, an area, or even an entire plant or facility, and examines it in context with all of the downtime. It may be common at some plants or facilities to refer to breakdowns as unplanned downtime. The total downtime should be all lost time, whether it is due to maintenance, operations, purchasing, transportation, or even an external supplier.

• *Strengths*

This indicator identifies whether the breakdown or unplanned downtime is actually a problem at the plant or facility. It may be that downtime is caused by another problem, rather than the preventive maintenance program. If that is the case, then this indicator highlights the fact.

• *Weaknesses*

The largest weaknesses to using this indicator are the proper classification of downtime and the accurate record keeping required. Downtime must be closely tracked and categorized. If an equipment-related breakdown is not closely tracked, then the time the operator is taking a break, procuring raw materials, or even eating lunch may be included in the breakdown time. This inflates the downtime and obscures other problems. Unless accurate records are kept, the breakdown downtime becomes a "catch all" and is not useful as a management tool.

2. Emergency Man-Hours

This indicator highlights the resources being allocated to plant or facility breakdowns. When the level of resource consumption for emergency or breakdown activities is high, then the productivity rates for the labor resources, whether in house or contract, are low. This indicator may be used at a department, area, or even a plant level. It may also be useful to examine it by trade or craft line. This indicator is also useful for examining work distribution. A typical distribution would examine the resources used in at least 4 categories:

• Preventive Maintenance
• Emergency or Breakdown Maintenance
• Repair or Corrective Maintenance
• Routine (or standing) Maintenance

Man-Hours Spent on Emergency Jobs
  Total Man-Hours Worked   Expressed as a %

This indicator takes the time spent on emergency or breakdown work and divides it by the total man hours expended. This indicator is then expressed as a percentage. It should examine total resources, not just maintenance. If there are operators involved, or contractors, their time should be included also.

- *Strengths*
    This indicator is useful for examining if maintenance labor is being consumed by emergency or breakdown work. Typically, if the amount of emergency or breakdown work consumes more than 20% of the maintenance labor resource, then the preventive maintenance program is viewed as ineffective. This indicator, then becomes a key to preventive maintenance program evaluation.

- *Weaknesses*
    This indicator, as with almost all, is dependent on accurate data collection. Without accuracy, then a problem with the preventive maintenance may go undetected. Additionally, what is classified as an emergency or breakdown may need clarification.

3. Cost of Breakdown Repairs

This indicator examines breakdowns in yet another way: the cost of the repairs. This would include the cost of the labor, materials, rental equipment, contractors, or any other direct maintenance cost. However, the cost of lost production or throughput should not be included in this calculation. This figure is then divided by the total maintenance cost and a percentage is derived. This indicator can again be calculated at different levels—the maintenance department level, at a trade or craft level, at a production department or areas level, or even at an equipment level.

Direct Cost of Breakdown Repairs
Total Direct Cost of Maintenance   Expressed as a %

This indicator is calculated by taking the direct cost of maintenance for all of the breakdown or emergency repairs and then dividing the cost by the total direct maintenance cost. The result is expressed as a percentage. Because the cost to perform maintenance in a reactive mode is considerably higher than the cost to perform maintenance in a planned mode (by as much as 2-to-4 times), this indicator will not match the percentage in the previous indicator.

- *Strengths*

    This indicator will highlight the impact the breakdown or emergency work is having on the maintenance budget. It can be used to cost justify improvements in the preventive maintenance program, when the percentage of maintenance dollars on break down or emergency activities is clearly shown.

- *Weaknesses*

    This indicator requires that all breakdown or emergency repairs be clearly identified. Even small activities, in the 5–10 minute range, must be clearly identified, or else many costs are not correctly identified. When the small activities are included, preventive maintenance problems are often exposed and then can easily be corrected.

4. Preventive Maintenance Compliance

This indicator examines the number of preventive maintenance tasks that are scheduled compared to the number of preventive maintenance tasks completed. This indicator is typically compiled on a weekly basis. It is useful for highlight a preventive maintenance program that may be developed, but not effective. The effectiveness is hampered by failure to complete the tasks that are scheduled. The reason may be that production is overcommitted and won't release the equipment for maintenance, or that the maintenance resources are overcommitted on breakdowns and emergency work and don't have the capacity to complete the scheduled preventive maintenance tasks.

Preventive Maintenance Tasks Completed
Preventive Maintenance Tasks Scheduled
Expressed as a percentage

This indicator takes the total number of preventive maintenance tasks completed (usually weekly) and divides them by the total number of preventive maintenance tasks scheduled. The result is then expressed as a percentage. The goal is to have 100% completion of the scheduled tasks. Although this number is not easily achieved, it should be the goal of all organizations. All preventive maintenance tasks should be included. This means tasks performed by maintenance, operations, or even contractors.

• *Strengths*

   This indicator is an effective measure of the compliance an organization has with its preventive maintenance program. It is one of the key indicators for any preventive maintenance program. If the indicator is graphed by week over a period of six months, it can be correlated with the percentage of maintenance activities that are breakdown or emergency. It will show that as completion rate goes up, the breakdowns and emergencies go down. Conversely, as the completion rate drops, the breakdowns and emergencies increase. If accurately tracked, the correlation is undeniable, and can be used to gain support for the preventive maintenance program.

• *Weaknesses*

   The weakness highlighted is not for the indicator, but for a type of preventive maintenance schedule that obscures or hides the fact preventive maintenance tasks are not completed. This is the dynamic or sliding preventive maintenance schedule. The weakness is that the preventive maintenance task is based on last completion date, not on a fixed schedule or a usage counter. This means that if the task is not completed this month, it is not rescheduled until it is completed. There are actually cases where monthly tasks have not been completed for three-to-six months and they do not show up as late or overdue. It is not recommended that any organizations truly serious about their preventive maintenance program ever use sliding or dynamic schedules.

5. Preventive Maintenance Estimates Compliance

   This indicator examines the estimates of labor and materials on the preventive maintenance task plans and compares them with the actual costs of performing the task. This highlights the accuracy of the estimates. If the estimates are inaccurate, then adjustments can be made and the accuracy can be insured. This indicator is particularly vital when the maintenance organization is using a scheduling system that is integrated with the production scheduling system. Inaccuracies in such a system will have dramatic negative consequences over time.

$$\frac{\text{Estimated PM Task Cost}}{\text{Actual PM Task Costs}} \qquad \text{Expressed as a percentage}$$

Note: This can be on a weekly basis

As can be seen by the formula, the indicator is calculated by dividing the estimated cost by the actual cost. The result is expressed as a percentage. A caution must be highlighted: this indicator should not be measured over a small window of time. On occasion, it is possible for a task to exceed the estimated cost due to exposed problems. However, if this analysis is performed on a semi-annual or annual basis, the results should provide a good indication of the accuracy of the estimates.

• *Strengths*
The strengths of this indicator include the ability to effectively monitor the accuracy of the preventive maintenance task estimates. If the accuracy of the individual estimates are not constantly monitored, then the overall accuracy of the estimated labor and materials required to perform the preventive maintenance tasks will be inaccurate, leading to budgetary problems.

• *Weaknesses*
The biggest weakness to this indicator is the issue of charging non-preventive maintenance tasks to the preventive maintenance activities. For example, if a problem uncovered during the performance of the task is corrected while performing the task, how is the additional labor and material charge billed? If the actual charges, which are repairs, are charged to the preventive maintenance work order, then the estimate is exceeded and the integrity of the estimate is in question. It is a good practice to complete the repair and then charge the costs to a new work order written to identify the work that was actually performed.

6. Breakdowns Caused by Poor PMs

This indicator examines the root causes of breakdowns and then investigates whether the root cause should have been detected as part of the preventive maintenance program. This provides an evaluation of the effectiveness of the preventive maintenance task and also the thoroughness of the individual carrying out the task. For example, lubrication-related failures should not occur on equipment that is inspected and lubricated as part of the preventive maintenance program. The breakdown indicates a failure of the preventive maintenance program. Modifications to the task listing, the retraining of an individual, or the addition of some visual control technique may be required to insure there is no repetition of the failure.

$$\frac{\text{Breakdowns Caused by Items that should have been}}{\text{Inspected, Serviced, or a Part of the PM Program}}{\text{Total Number of Breakdowns}}$$
Expressed as a percentage

The formula indicates that the total number of breakdowns that could have been prevented or detected by the preventive maintenance program is compared to the total number of breakdowns. The resulting percentage indicates the opportunity for improvement for upgrading or changing the preventive maintenance program. An additional driver for improvement can be uncovered if the losses (maintenance costs, equipment damage, downtime costs) incurred by the breakdowns are also included.

- *Strengths*
  This indicator is beneficial to any organization desiring to improve their preventive maintenance program. It provides an accurate insight into the effect that preventive maintenance is having on the equipment breakdowns. Monitoring this indicator helps to insure that the preventive maintenance policy is cost effective.

- *Weaknesses*
  The greatest weakness to using this indicator is procedural. In other words, the organization must be committed to completing accurate and detailed root cause analysis of equipment breakdowns. If the root cause is merely assumed or guessed, then the true effectiveness of the preventive maintenance program is obscured.

7. Preventive Maintenance Efficiency
  This indicator examines the amount of work that is generated from the preventive maintenance program. When carrying out the preventive maintenance inspection, the inspector will uncover components or systems showing signs of wear, or even an impending failure. The inspector will then write work orders to correct the problem before a breakdown occurs. This may involve adjustments, changing components, or even a major overhaul. The point is, some work should be generated from the inspections and service; otherwise it is quite likely that the preventive

maintenance tasks are being carried out too frequently.

Total Number of Work Orders Generated from PM Inspections
Total Number of Work Orders Generated
Expressed as a percentage

The formula shows the efficiency is measured by taking the total number of work orders generated from the preventive maintenance program and dividing this by the total number of work orders submitted. This is generally examined on a monthly basis, although other frequencies can be acceptable, depending on the inspection frequency. The resulting percentage will highlight whether the preventive maintenance program is effective in finding developing equipment problems proactively.

- *Strengths*
  This indicator is effective for preventive maintenance program evaluations. It is viewed as effective if the majority of the work orders submitted are found in performance of the preventive maintenance program. Although this may appear to be performing too much preventive maintenance, this fact will not be established until further factors are applied in the reliability centered maintenance approach.

- *Weaknesses*
  This indicator can be misleading if excessive work is performed by the inspectors while actually carrying out the preventive maintenance task. If, rather than having a work order written, the repairs are hidden in the preventive maintenance charges, the effectiveness of the inspections will be called into question. A second problem may be motivating individuals to fill out the necessary documentation to establish the data. If the inspectors would rather do the task than write up a work order, the amount of work discovered by the preventive maintenance program is hidden. Also contributing to this situation may be the lack of good inspection skills. Do the inspectors really know how to inspect, what to look for, and how to find true root causes? These questions can be resolved by good testing and training of the inspectors.

8. Equipment Uptime

This indicator is used to highlight the amount of uptime that is required for the equipment to meet the production forecasts. In a sold-out market condition, the required uptime may be 100%. However, if the equipment is required 100% of the time because the plant has to continually make up production losses caused by unreliable equipment, then, as the preventive maintenance program becomes more effective, the desired uptime may change. Requiring 100% uptime makes it difficult to perform the right level of maintenance on the equipment. This contributes to future problems. This indicator may help determine if the organization has a realistic expectation of the equipment output. If 100% is expected all of the time, the organization's technical understanding and commitment to the longevity of its assets must be questioned.

Desired Equipment Uptime – Downtime
Desired Equipment Uptime   Expressed as a percentage

The indicator is the desired uptime minus the downtime divided by the desired uptime. The result is expressed as a percentage. Some organizations refer to this as availability. The percentage should be evaluated on a weekly or monthly basis and trended. A decrease in uptime will indicate a problem with the preventive maintenance program, possibly indicating a change in the life cycle phase of the equipment. It can also indicate a possible change in operational schedules, which severely impact calendar based preventive maintenance programs.

• *Strengths*
This is a good indicator because ultimately the preventive maintenance program is designed to maximize uptime. Most of the information to calculate the uptime comes from the production or facilities groups. Using this calculation helps to insure their understanding and support of the preventive maintenance program. This indicator may be superior to the one used previously because all downtime is included. This will foster more departmental support for the indicator because any downtime related to causes within their control is also exposed, not just maintenance.

- *Weaknesses*
  The weakness to this indicator is that all causes of downtime are tracked and used in the calculation. This makes the indicator different than the previous one, where just the maintenance-related downtime was tracked. This differentiation requires very accurate data so that the preventive maintenance program does not get blamed for downtime it can not prevent.

9. Number of PMs Overdue as a Percentage

This indicator examines the number of preventive maintenance tasks that are not being completed on schedule. This indicator is valuable for spotting trends where compliance to schedule is beginning to slip. This indicator will forecast problems because once the schedule begins to slip, the breakdown or emergency requests will begin to rise soon after. Paying attention to this indicator will allow a proactive approach to managing the preventive maintenance program to be enforced. This indicator is most effective when monitored on a weekly basis and then trended over a rolling six-month time period.

> Number of PMs Overdue
> Total Number of PM's Outstanding   Expressed as a Percentage

As can be seen by the formula, this indicator is the number of preventive maintenance tasks that are outside their due time period divided by the number of preventive maintenance tasks that are currently in the active backlog. This gives an accurate representation of the level of effort required to keep the preventive maintenance program in compliance. The goal, of course, is to keep this percentage as low as possible.

- *Strengths*
  This is a required indicator for any company monitoring the progress of their preventive maintenance program. Without this indicator, there would be virtually no way to track the compliance status of the preventive maintenance program.

- *Weaknesses*
  The only weakness of this indicator is the challenge to keep the data accurate. There are organizations that cancel preventive maintenance tasks so they don't clutter the work order backlog.

This practice is not recommended because it can alter the accuracy of the preventive maintenance program compliance data. It is better to allow the problems to be exposed so that they can be corrected than to hide the problems by manipulating the work order data.

10. Percentage of Overtime

This indicator is not always directly an indicator of the preventive maintenance program effectiveness. However, it can be impacted, so it is good to mention it at this point. In many organizations, overtime is worked in response to equipment breakdowns or emergency work. Where the overtime rates are high, it can be an indicator of the ineffectiveness of the preventive maintenance program.

Hours Worked as Overtime
Total Hours Worked   expressed as a percentage

As the formula shows, the indicator is derived by dividing the hours worked as overtime by the total hours worked. This percentage then shows the premium time that is being used to perform work. Proactive maintenance organizations work 5% or less of their total time as premium time.

Preventive Maintenance Program Problems

The following are the most common disablers for preventive maintenance programs. If these conditions exist in an organization, it will be difficult if not impossible for a sustainable preventive maintenance program to be implemented.

1. Lack of Management Support

This is the most critical single factor to the success or failure of a preventive maintenance program. If management is not committed to a preventive maintenance program, it will fail. This condition leads to the sub-optimization of all of the rest of the maintenance initiatives within an organization.

Although there is no magic answer to solving this problem, companies that have management support obtain it with financial justification. The fact is that preventive maintenance can address many issues that affect a company's ability to remain profitable. Some of the issues that are impacted by preventive maintenance include:

ISO, OSHA, EPA, PSM, etc.
Most regulatory programs require equipment that a) is safe to operate and maintain, b) is not hazardous to the environment, and c) can hold specifications to produce a quality product.

Total Quality Management
The support for this program is clearly seen in the ISO-9000 and QS-9000 program requirements for preventive maintenance and tracking of preventive maintenance compliance.

Just In Time
Simply, it is impossible to produce products on an exacting schedule without reliable and maintainable equipment. It is impossible to have reliable and maintainable equipment without an effective preventive maintenance program.

Customer Service Orientation
Again, simply put, it is impossible to produce the lowest cost product, with perfect quality, and deliver it in a timely manner to satisfy a company's customers without a preventive maintenance program to insure equipment reliability.

Capacity Constraints
Preventive maintenance insures not just uptime of the equipment, but also performance efficiency. This means that when the equipment is operating, it will produce at design capacity with the desired uptime. This performance level helps to insure that a company does not develop equipment-related capacity constraints.

Redundant Equipment
When equipment operates as designed when it is required to operate, it reduces the need for redundant equipment to back up unreliable equipment or to supplement the existing equipment capacity. This keeps the return on net assets indicator at a best practice level.

Energy Consumption

Well-maintained equipment requires 6% to as much as 11% less energy to operate than poorly-maintained equipment. When a company considers heat exchangers, coolers, HVAC systems, steam leaks, and air leaks, it can quickly see how energy consumption can be an area of considerable savings.

Usable Asset Life

Well-maintained equipment lasts 30–40% longer than poorly-maintained equipment. It is easy to develop a "don't maintain, just replace" attitude with equipment. This leads to unnecessary capital expenditures. It is easy to see if this is a problem for a company. Just examine how often equipment is replaced in kind—no major upgrade in engineering or technology, just replaced because it wore out. Is it possible that the purchase could have been deferred if proper preventive maintenance had been performed?

By examining these eight areas, it may be possible to convince the appropriate managers and obtain the management support necessary to have a successful preventive maintenance program.

2. Lack of Maintenance Skills

This area is developing into one of the major problems facing preventive maintenance programs today. The skills necessary to inspect and perform basic maintenance tasks on equipment today seem to be deteriorating, just routine tasks like:

- Proper lubrication of bearings
    The right lubricant
    The right quantity
    The right frequency
    The right application method

In many companies, basics such as these are virtually ignored. The problem compounds when proper installation and maintenance of basic components such as belts, chains, gears, and pneumatic and hydraulic systems are considered. It does little if any good to schedule preventive maintenance activities, if they can not be carried out correctly.

The answer to correct this problem is training (See Chapter 7) and enforcement of the learned behavior. This means that the knowledge will have to be made available to anyone required to operate or maintain the equipment. Once the knowledge reaches critical mass, it will require peer pressure and equipment ownership to assure continuance of good practices.

3. Wrong Equipment Selected

Selected the appropriate equipment is a preventive maintenance program start-up problem. It occurs when the equipment is selected to start the preventive maintenance program. When starting a preventive maintenance program, the mission critical equipment should be selected. It is imperative that results be shown early in the preventive maintenance program. The equipment selected should be constraint equipment, equipment causing a bottleneck, or equipment that has no backup and will severely impact production or availability of the facility if it incurs downtime.

This problem can be overcome by prioritizing the equipment when beginning the preventive maintenance program. The equipment selected should meet the previously-mentioned criteria. It may be best to poll different department, managers, and shop floor personnel to insure that support for the equipment selected is developed. Thus, an attitude fostering organizational support for the preventive maintenance program is developed.

4. Not Changing / Updating PMs

This problem develops after the preventive maintenance program has been in place for a time period. The preventive maintenance program was probably effective, and then the level of breakdowns and the amount of reactive maintenance both started to increase. Even though the preventive maintenance program is in compliance, results start diminishing.

This is due to the fact that the equipment is entering a different phase of its life cycle. What may have been the right level of service and activities is in the past; the equipment maintenance needs are changing as the equipment is aging. The preventive maintenance tasks should be re-evaluated in light of the current equipment problems. It may be that when the preventive maintenance tasks were developed, the daily, weekly, and monthly tasks were defined. However, the service required at semiannual, annual, and biannual frequencies may not ever have been developed. This lack of needed service allows components on the equipment to develop undetected problems and fail.

The preventive maintenance tasks must be evaluated in context with the long-term equipment needs. This insures that preventive maintenance occurs for the entire life cycle of the equipment.

5. Poor Schedule Compliance

This problem occurs for several reasons, but always impacts the effectiveness of the preventive maintenance program. When tasks are scheduled and not completed within the assigned time frame, the equipment begins to deteriorate. Although the equipment may not begin to fail immediately, it begins to develop multiple deteriorated components. The interaction of the worn component conditions begins to mask problems. Operating the equipment becomes more complicated, as it no longer stays in adjustment or hold specifications. The equipment also will no longer run at design speed, but the deterioration requires it to be slowed 10% or more, reducing its capacity. Troubleshooting the equipment also becomes more difficult as one problem leads to another. The equipment will require a rebuild to reach an acceptable baseline, where the preventive maintenance tasks will once again be effective.

The only cure is to dedicate the resources and make the release time to keep the preventive maintenance program in compliance. This may require management support because production schedules may have to be altered or over time may have to be authorized.

6. Insufficient Detail on PM Sheets

This is typically a start -p problem with the preventive maintenance program. The proper level of detail is not developed, so items are missed on the preventive maintenance inspections or services. Some samples of poor detail include:

- Check the motor to see if t is Hot
- Check the belt drive
- Check the chain drive

These are examples of vague preventive maintenance inspections. For example, for the motor, how hot is hot? The task description should contain temperatures, pressure setting, flow values, etc.

Some may argue that that level of detail is expensive and time consuming. In itself, that statement is true. However, what does a missed inspection point cost when the equipment fails? The lack of preventive maintenance inspection details allows for items to be overlooked or viewed incorrectly. This contributes to breakdowns and overall loss of preventive maintenance efficiency.

One of the largest contributing causes to the lack of detail is the lack of resources during the initial development of the preventive maintenance program. For example, how can the resources be calculated? Using the following information may be insightful:

Number of equipment items -	1,000
Average number of PMs per item	x3
Total PMs required	3,000
Average 1 hour per PM	
For development	x1
Total man-hours required	3,000 or 1.5 man-years of effort

Because most companies do not allocate this level of resource, the preventive maintenance tasks are partially developed, with the hope of going back someday and finishing the development. But, in most organizations, this never happens. Therefore, the preventive maintenance program is ineffective.

The only solution is first to dedicate the resources necessary to develop the initial task details, then to realize the full benefits of preventive maintenance.

7. PM Data Not Being Recorded

This problem occurs after the preventive maintenance program is implemented and the completed inspections are turned in for processing. The problem is, the inspections are never reviewed and the notations are not permanently documented. The cause of the problem is usually that there are no resources to transfer the inspection results to a database so they can be analyzed. As a result, the comments made by the inspectors are lost. Any subsequent work that was to be requested and performed is also lost. This prevents evaluation of the preventive maintenance program results and effectiveness. As before, the preventive maintenance program will deteriorate as the equipment ages, not being able to change to meet the equipment's changing needs.

The lack of data gather is also commonly caused by the lack of an easy to use computerized maintenance management system. In many organizations, resources to record data are scarce. Any CMMS used to collect data must make it easy to input and then analyze.

This problem has a two-part solution. First, the correct staffing level must be determined, based on the amount of data collection and analysis the organization is performing. When understaffed, the data accuracy ultimately suffers and what is collected is of no value. Once the correct staffing level is provided, the organization should plan to utilize the most effective CMMS it can afford. This reduces the amount of frustration that employees have when recording and analyzing data on antiquated software.

8. Lack of Understanding of EPA, OSHA, ISO Regulations

This issue is actually a lack of education and the disciplines to carry out data recording based on that education. The regulatory requirements for maintenance organizations are complex and require extensive training to accurately understand. The issue is that most maintenance management positions are high turnover positions, with newer and younger replacements being hired.

This problem places the company under tremendous pressure. If the preventive maintenance is not carried out on the equipment, the company may be in violation of a regulatory standard. However, the maintenance manager may not even have a good understand of which preventive maintenance activities require priority. This leads to compliance issues and ultimately a failure of the preventive maintenance program.

The only solution to this problem is a) effective education for the maintenance supervision on regulatory requirements and then b) the ability to enforce the requirements to collect the data and keep the company equipment in compliance.

APPENDIX A

SAMPLE OPERATOR INSPECTIONS

V-Belt Drive

Record the information from the inspection in the appropriate column. Check any required actions.

Additional Description	Good Condition	Requires Lubrication	Requires Adjustment	Requires Replacement	Requires Cleaning	Has Unusual Vibration	Excessive Temperature	Requires Tightening	Work Order Required	See Additional Comments
1. Electrical Motor										
a. Bearings										
b. Base and Bolts										
c. Temperature — Should be less than 120 degrees F										
d. Vibration										
e. Noise										
2. V-Belt										
a. Alignment										
b. Sheave Condition										
c. Belt Condition										
3. Gearcase										
a. Gears										
b. Bearings										
c. Base and Bolts										
4. Bearings										
a. Seals — Evidenced by excessive motion or play										
b. Excessive Wear										
Additional Comments										
Work Order Written:										

Figure A-1 Belt Drive

Chain Conveyor

Record the information from the inspection in the appropriate column. Check any required actions.

	Additional Description	Good Condition	Requires Lubrication	Requires Adjustment	Requires Replacement	Requires Cleaning	Has Unusual Vibration	Excessive Temperature	Requires Tightening	Work Order Required	See Additional Comments
1. Electrical Motor											
a. Bearings											
b. Base and Bolts											
c. Temperature	Should be less than 120 degrees F										
d. Vibration											
e. Noise											
2. Couplings											
a. Alignment											
b. Lubrication											
3. Gearcase											
a. Gears											
b. Bearings											
c. Base and Bolts											
4. Bearings											
a. Seals											
b. Excessive Wear	Evidenced by excessive motion or play										
5. Chain											
a. General Condition											
b. Lubrication											
Additional Comments											

Work Order Written:

#

#

#

Figure A-2 Chain Conveyor

Chain Drive

Record the information from the inspection in the appropriate column. Check any required actions.

	Additional Description	Good Condition	Requires Lubrication	Requires Adjustment	Requires Replacement	Requires Cleaning	Has Unusual Vibration	Excessive Temperature	Requires Tightening	Work Order Required	See Additional Comments
1. Electrical Motor											
a. Bearings											
b. Base and Bolts											
c. Temperature	Should be less than 120 degrees F										
d. Vibration											
e. Noise											
2. Couplings											
a. Alignment											
b. Lubrication											
3. Gearcase											
a. Gears											
b. Bearings											
c. Base and Bolts											
4. Bearings											
a. Seals	Evidenced by excessive motion or play										
b. Excessive Wear											
5. Chain											
a. Chain Condition											
b. Sprocket Condition											
Additional Comments											
Work Order Written:											
#											
#											
#											

Figure A-3 Chain Drive

Belt Conveyor Inspection

Record the information from the inspection in the appropriate column. Check any required actions.

	Additional Description	Good Condition	Requires Lubrication	Requires Adjustment	Requires Replacement	Requires Cleaning	Has Unusual Vibration	Excessive Temperature	Requires Tightening	Work Order Required	See Additional Comments
1. Electrical Motor											
a. Bearings											
b. Base and Belts											
c. Temperature	Should be less than 120 degrees F										
d. Vibration											
e. Noise	Anything Unusual										
2. Couplings											
a. Alignment											
b. Lubrication											
3. Gearcase											
a. Gears											
b. Bearings											
c. Base and Bolts											
4. Bearings											
a. Seals	Evidenced by excessive motion or play										
b. Excessive Wear											

Additional Comments

Work Order Written:

Figure A-4 Conveyor

Gear Case

Record the information from the inspection in the appropriate column. Check any required actions.

	Additional Description	Good Condition	Requires Lubrication	Requires Adjustment	Requires Replacement	Requires Cleaning	Has Unusual Vibration	Excessive Temperature	Requires Tightening	Work Order Required	See Additional Comments
1. Electrical Motor											
a. Bearings											
b. Base and Bolts											
c. Temperature	Should be less than 120 degrees F										
d. Vibration											
e. Noise											
2. Couplings											
a. Alignment											
b. Lubrication											
3. Gearcase											
a. Gears											
b. Bearings											
c. Base and Bolts											
4. Bearings											
a. Seals											
b. Excessive Wear	Evidenced by excessive motion or play										

Additional Comments

Work Order Written:

#

#

#

Figure A-5 Gear Case

Hydraulic Lift Record the information from the inspection in the appropriate column. Check any required actions.	Additional Description	Good Condition	Requires Lubrication	Requires Adjustment	Requires Replacement	Requires Cleaning	Has Unusual Vibration	Excessive Temperature	Requires Tightening	Work Order Required	See Additional Comments
1. Electrical Motor											
a. Bearings											
b. Base and Bolts											
c. Temperature	Should be less than 120 degrees F										
d. Vibration											
e. Noise											
2. Couplings											
a. Alignment											
b. Lubrication											
3. Hydraulic System											
a. Piping											
b. Valves											
c. Leaks											
d. Cylinder											
Additional Comments											
Work Order Written:											
#											
#											
#											

Figure A-6 Hydraulic Lift

Hydraulic Motor

Record the information from the inspection in the appropriate column. Check any required actions.

	Additional Description	Good Condition	Requires Lubrication	Requires Adjustment	Requires Replacement	Requires Cleaning	Has Unusual Vibration	Excessive Temperature	Requires Tightening	Work Order Required	See Additional Comments
1. Electrical Motor											
a. Bearings											
b. Base and Bolts											
c. Temperature	Should be less than 120 degrees F										
d. Vibration											
e. Noise											
2. Couplings											
a. Alignment											
b. Lubrication											
3. Hydraulic System											
a. Motor											
b. Piping											
c. Leaks											
d. Valves											
a. Seals											

Additional Comments

Work Order Written:

#

#

#

Figure A-7 Hydraulic Motor

Pneumatic Cylinder

Record the information from the inspection in the appropriate column. Check any required actions.

Additional Description	Good Condition	Requires Lubrication	Requires Adjustment	Requires Replacement	Requires Cleaning	Has Unusual Vibration	Excesive Temperature	Requires Tightening	Work Order Required	See Additional Comments
1. Compressor										
a. Bearings										
b. Base and Bolts										
c. Temperature										
c. Vibration										
e. Noise										
2. Couplings										
a. Alignment										
b. Lubrication										
3. Pneumatic System										
a. Piping, Hoses										
b. Leaks										
c. Valves										
c. Motor										

Additional Comments

Work Order Written:

#

#

#

Figure A-8 Pneumatic Cylinder

Pneumatic Motor

Record the information from the inspection in the appropriate column. Check any required actions.

Additional Description	Good Condition	Requires Lubrication	Requires Adjustment	Requires Replacement	Requires Cleaning	Has Unusual Vibration	Excessive Temperature	Requires Tightening	Work Order Required	See Additional Comments
1. Compressor										
a. Bearings										
b. Base and Bolts										
c. Temperature										
d. Vibration										
e. Noise										
2. Couplings										
a. Alignment										
b. Lubrication										
3. Pneumatic System										
a. Piping, Hoses										
b. Leaks										
c. Valves										
d. Motor										
Additional Comments										
Work Order Written:										
#										
#										
#										

Figure A-9 Pneumatic Motor

Appendix B

Developing System PM Inspections

Inspection 1: Typical Motor and Gearcase Combination

Refer to Figure B-1 for this section.

Electric Motor

Item 1 in the figure is the electric motor. The most common inspection on the motor during operation is for heat. The motor should be approximately 20 to 25°F higher in temperature than its surrounding environment. If the motor is higher in temperature than this, its life will be shortened due to the effect the heat has on the insulation. For every additional 20°-temperature rise, the life of the insulation will be cut in half. When the insulation fails, so does the motor. If the temperature of the motor becomes too high, efforts should be made to find the problem, so the motor can be cooled back down.

1A denotes the bearings that support the rotating part of the motor (usually the rotor or the armature). The inspection of the bearings should

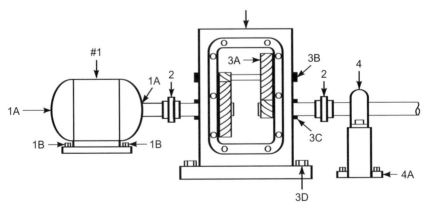

Figure B-1. For inspection 1: typical motor and gearcase.

consider three basic factors: heat, noise, and vibration. These three indicators of trouble may appear singly or in any combination. Monitoring these three conditions may require nothing more than the use of your normal senses. If the equipment is critical in nature, it may require the use of some monitoring or nondestructive testing equipment. For measuring the temperature, some form of hand-held thermometer may be used. If the bearings are in a difficult-to-reach or unsafe location, a permanent monitor may be used. If vibration is to be checked, the use of a hand-held meter may be sufficient. These instruments will be able to indicate such conditions as insufficient lubrication, misalignment, contamination, or normal wear of the bearings. If any of these conditions are detected, measures should be taken to correct the problem as soon as it can be scheduled.

1B denotes the base bolts of the motor. The base bolts should always be checked for the correct torque. If the bolts aren't tightened correctly, they may come loose during the operation of the equipment, resulting in unwanted downtime of the equipment. Also, during removal or installation of the motor, the base should be checked to be sure that it's level and in good shape, to avoid putting additional stress on the motor when it's bolted into place.

Couplings

Item 2 in Figure B-1 is a coupling, which is used to connect twoshafts together. The most critical item is the alignment of the couplings. If they aren't aligned to within ±0.005, the coupling will have a shortened service life. While 0.005 is used as a general figure, the closer the alignment is to perfect, the longer the life of the coupling will be. In addition to shortening the life of the coupling, misalignment will also shorten the life of the bearings and related components in the drive. In addition to alignment, certain couplings require lubrication. It is important to provide the correct amount of the correct lubricant to these couplings. If sufficient lubricant isn't provided, metal-to-metal contact inside the coupling will occur, resulting in a reddish brown color to the lubricant and a bluish color to the coupling teeth. Too much lubricant will result in overheating of the coupling due to fluid friction. The extreme heat will then destroy the lubricant, resulting in failure of the coupling.

Gearcase

Item 3 is the gearcase. The gearcase has one important consideration - the lubricant. It must be maintained at the correct level within the gearcase. An insufficient level of lubricant allows metal-tometal contact between the rotating parts of the gearcase resulting in rapid destruction of all related components within the gearcase.

Item 3A is the gearing. The gears need proper lubrication at all times. If the gears are left to run without lubrication, the tooth surfaces will be destroyed very quickly, resulting in failure of the drive. The tooth surfaces can be inspected periodically for any unusual wear patterns, which may indicate some problem within the case such as contaminated lubricant, misalignment of the gears, insufficient backlash, overload of the gearcase, or pitting of the teeth. Visual inspections should be made semi-annually in critical applications, and annually in general applications.

Items 3B and 3C are the bearings supporting the shafts of the gearcase. The most important application with the bearings is the lubricant. The lubrication method may be splash or spray, but it must get to the bearings, or failure of the bearings will result. While this is bad enough, the failure of the bearings can also result in changes of the internal geometries of the gearcase. This will allow damage to the more expensive gears to occur and may result in a more costly and time-consuming breakdown than would have occurred if the bearings had been properly lubricated. It's very important to detect any problems with the shaft bearings in a gearcase before they occur.

3D denotes the basebolts for the gearcase. It's important that these bolts always be tightened to the correct torque values. If the case comes loose during operation, damage will occur to the bearings, gearing, and the couplings. Also, anytime the gearcase is removed, the base should be inspected to ensure that no defect develops that would allow a breakdown during operation.

Bearings

Item #4 is a bearing that supports the end of the drive train. This bearing is very important because it carries the output of the gearcase to its final destination. The inspector should watch for heat, vibration, and noise. If these conditions are detected, it's important to correct the problem before a failure occurs.

4A denotes the mounting or base bolts for the bearing. The same consideration should be given them as was given the base bolts on the motor and gearcase.

Preventive Maintenance Inspection
Inspection 1: Typical Motor and Gearcase Combination

Check the box or boxes on the right that describe the condition of the units named in the left column.	O.K.	Requires lubrication	Requires adjustment	Requires replacement	Requires cleaning	Excessive vibration	Excessive heat	Loose	See additional comments
1. Electric Motor									
A. Bearings									
B. Base and bolts									
C. Temperature									
D. Vibration									
E. Noise									
2. Couplings									
A. Alignment									
B. Lubrication									
3. Gearcase									
A. Gears									
B. Bearings									
C. Bearings									
D. Base and base bolts									
4. Bearings									
A. Base and bolts									
B. Excessive play or motion									

Additional Comments:

Inspection 2: Typical Belt Drive

Refer to Figure B-2 for this section.

Electric Motor

Item 1 is the electric motor. The most common inspection item for the motor is for heat. The motor should be approximately 25 to 29°F high-

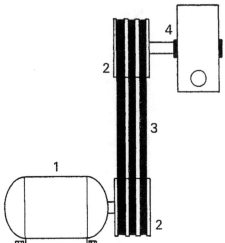

Figure B-2. For inspection 2: typical belt drive.

er in temperature than the surrounding environment. If the motor's temperature is higher than this, it will have a shorter life due to the deteriorating effect the heat has on the insulation of the motor. For every 20°-temperature rise above the environmental temperature, the life of the insulation is cut in half. When the insulation fails, so does the motor. If the temperature rises above this level, efforts should be made to find the problem, so the motor can be cooled back down again.

Another inspection point on the motor is the bearings. The bearings support the rotating part of the motor (usually called the anna- ture or the rotor). The inspection of the bearings should include three basic items: heat, noise, and vibration. These three indicators of trouble may appear singly or in any combination. Monitoring these three conditions may require nothing more than the use of the inspector's natural senses. If the equipment is critical in the manufacturing process, it may be advisable to use some form of monitoring or nondestructive testing equipment. For monitoring the temperature, some form of hand-held thermometer may be sufficient. If the bearings are in a difficult-to-reach or unsafe area, some form of temperature-monitoring device may be used. Vibration may be measured by a hand-held meter or a permanent monitoring device. Either of these types of vibration meters can determine if conditions such as insufficient lubrication, misalignment, contamination, or normal wear are occurring in the bearing. If any of these conditions are detected, steps should he taken to correct the problem. If they're not corrected, then the bearing will fail, and the replacement will also fail quickly.

Another item on the motor to inspect is the base. The base bolts should always he checked to ensure that they're tightened to the correct torque specifications. If they're not, they may come loose during operation, allowing the motor to misalign with the remaining parts of the drive. Periodically, the base of the motor and the mount- ing for the motor should be inspected. The foundation may develop faults, allowing the motor to loosen on the base. Also, settling of the base may occur, allowing it to place excessive stress on the motor base. The base should always be completely clean when installing the motor.

Sheaves

Items labeled 2 are the sheaves - the grooves in which the belts run. The contact between the side of the belt and the grooves is the area that transmits the power. It is this area that must be inspected for wear. The sidewall gauge should be used to determine if wear exceeds allowable limits. If it does, the sheaves should be changed to prevent excessive belt wear. The sheaves should also be cleaned periodically with a stiff brush to remove any dirt or deposits that would eventually cause wear on the sides of the sheave. The area where the sheave mounts on the shaft should also be checked to determine if any looseness or excessive wear is occurring. This condition would allow the sheave to slip under heavy loads. If looseness has been observed, corrective action should be taken before a breakdown occurs during operation.

The alignment of the sheaves should also be examined. The sheaves should be in line to prevent excessive wear on the sheave sidewall and the belt. The more exact the alignment is made, the longer the drive life will be. Any time spent aligning the sheaves is time well spent.

Belt

Item 3 in Figure B-2 is the belt. Belt inspections can be merely observations of any unusual wear patterns on the belt. If the belt's surface appears to be normal, with no contamination or cracking apparent, the drive will usually be in relatively good condition.

Heat is a deadly enemy of belt drives. If the belt is exposed to heat, it overcures, causing it to become brittle and crack. Heat can be ambient, or may develop within the drive. Any slippage in the drive causes heat buildup, causing rapid deterioration. This points to the need for proper tension of the belt to eliminate slippage. The best tensioning method is the tension tool, which can be provided by any distributor. Because a belt can slip up to 20% before it makes any noise, sound shouldn't be used to determine if a belt is slipping.

Another test for slippage is the use of a strobe gun. A strobe light "freezes" the belt and sheave during operation. If the belt is slipping, it quickly shows under the light. If the belts exhibit any unusual tracking tendencies, consideration should be given to replacement. This would include running off the sheaves, or turning over in the sheaves. Both of these conditions indicate that some tension members are broken and the belt won't carry its rated load.

Bearings

Item 4 indicates the bearings on the drive. The reason they're included is to point out the effect excessive belt tension or too large a sheave may have on the gearing. If the belt is too tight, it puts an excessive load on the bearings, causing high temperatures, rapid wear, and premature failure. If the drive bearings exhibit these signs, their ratings should be checked, along with the belt tension and the weight of the sheaves.

Inspection 3: Typical Chain Drive

Refer to Figure B-3 for this section.

Electric Motor

Item 1 is the electric motor. The most common inspection item for the motor is for heat. The motor should be approximately 25 to 29°F high-

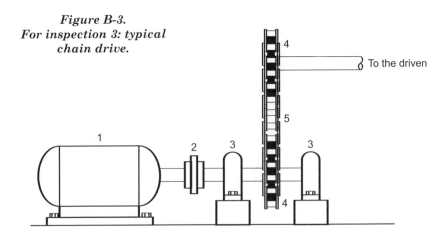

Figure B-3.
For inspection 3: typical
chain drive.

Preventive Maintenance Inspection
Inspection 2: Belt Inspection

Check the column that indicates the condition of the unit or what problem exists named in the left column.	O.K.	Requires lubrication	Requires adjustment	Requires replacement	Requires cleaning	Excessive vibration	Excessive heat	Loose	See additional comments
1. Electric Motor									
A. Bearings									
B. Base and bolts									
C. Temperature									
D. Vibration									
E. Noise									
2. Sheave									
A. Sidewall wear									
B. Dirt									
C. Shaft mounting									
d. Alignment									
3. Belt									
A. Cover wear									
B. Tension									
C. Tracking									
4. Bearings									
A. Temperature									
B. Noise									
C. Vibration									
D. Excess play or motion									

Additional Comments:

er in temperature than the surrounding environment. If the motor's temperature is higher than this, it will have a shorter life due to the deteriorating effect heat has on the insulation of the motor. Every 20°-temperature rise above the environmental temperature cuts the life of the insulation by one half. When the insulation fails, so does the motor. If the temperature rises above this level, efforts should be made to find the problem, so the motor can be cooled back down again.

Another inspection point on the motor is the bearings. The bearings support the rotating part of the motor (usually called the armature or the rotor). The inspection of the bearings should include three basic items: heat, noise, and vibration. These three indicators of trouble may appear singly or in any combination. Monitoring these three conditions may require nothing more than the use of the inspector's natural senses. If the equipment is critical in the manufacturing process, it may he advisable to use some form of monitoring or nondestructive testing equipment. For monitoring the temperature, some form of hand-held thermometer may be sufficient. If the equipment is in a difficult-to-reach or unsafe area, some form of temperature-monitoring device may be used. Vibration may be measured by a hand-held meter or a permanent monitoring device. Either of these types of vibration meters will be able to determine if conditions such as insufficient lubrication, misalignment, contamination, or normal wear are occurring in the bearing. If any of these conditions are detected, steps should be taken to correct the problem. If they're not corrected, the hearing will fail, and the replacement will also fail quickly.

Another item on the motor to inspect is the base. The base bolts should always be checked to ensure that they are tightened to the correct torque specifications. If they're not, they may come loose during operation, allowing the motor to misalign with the remaining parts of the drive. Periodically, the base of the motor and the mounting for the motor should he inspected. The foundation may develop faults, allowing the motor to loosen on the base. Also, settling of the base may occur, causing excessive stress on the motor base. The base should always he completely clean when installing the motor.

Couplings

Item 2 in Figure B-3 is the coupling, which is used to connect two shafts together. The most critical item is the alignment of the coupling halves. If they're not aligned within very close tolerances, rapid wear will occur. A rule of thumb is that they must be within 0.005 of an inch. Rigid couplings must be in exact alignment. Flexible couplings may be able to withstand 0.005. The closer the coupling alignment is to being exact, the longer the coupling will last. Misalignment also damages related items in the drive, such as the bearings and shafts. Correct alignment of couplings cannot be overemphasized.

Lubrication is another prime consideration in coupling inspections. If correct lubrication isn't provided for flexible couplings, rapid wear and complete failure will result. If metal-to-metal contact occurs in the coupling, welding and tearing of the coupling material occurs. It's also important not to overlubricate. If overlubrication occurs, fluid friction also occurs, which builds up heat and destroys the lubricant, resulting in rapid wear of the coupling material and failure of the coupling. If a coupling is opened up for inspection and a reddish brown color is observed, it should be cleaned and inspected and then properly relubricated.

Support Bearings

Item 3 refers to support bearings for the sprocket and shaft. These bearings can be inspected by observing the three basic signs of bearing wear: noise, heat, and vibration. If any of these signs , are observed, a problem exists with the bearing. This type of bearing in the figure is commonly called a pillow block bearing. If the bearing appears to be in good condition, the lubrication should be checked. If it's at the proper level (covering half the lowest ball or roller if oillubricated, or one-third of the housing if grease-lubricated), then it should be safe to move on to the next item.

Chain Sprocket

Item 4 is the chain sprocket. If it's kept in good condition and well lubricated, a sprocket should outlast three or more replacement chains. Sprocket wear occurs on the teeth. If the teeth are becoming hook shaped, then the sprocket will begin rapid wear of the chain.

This will also cause the chain to hang in the sprocket causing overloads on the drive. If the teeth are showing wear on the sides, the sprockets are usually out of alignment. If the alignment isn't corrected, the life of the drive is dramatically shortened.

Chain

Item 5 is the chain. The following inspection points can be applied to any type of chain drive, however, a roller chain drive is considered here. The primary inspection point is lubrication. Since a roller chain

Preventive Maintenance Inspection
Inspection 3: Chain Inspection

Check the column that indicates the condition of the unit or what problem exists named in the left column.	O.K.	Requires lubrication	Requires adjustment	Requires replacement	Requires cleaning	Excessive vibration	Excessive heat	Loose	See additional comments
1. Electric Motor									
A. Bearings									
B. Base and bolts									
C. Temperature									
D. Vibration									
E. Noise									
2. Coupling									
A. Alignment									
B. Lubrication									
3. Bearing									
A. Base and bolts									
B. Excessive play or motion									
4. Sprocket									
A. Tooth wear									
B. Lubrication									
C. Alignment									
D. Mounting to shaft									
5. Chain									
A. Elongation									
B. Side plate wear									
C. Tension									
D. Engagement of sprocket									

Additional Comments:

wears 300 times faster if it's run unlubricated than when well lubricated, lubrication is a primary concern. The lubricant should penetrate the chain joint to prevent metal-to-metal contact. A rule of thumb is to use a good 30-weight oil at normal temperatures, thinner oil in colder weather, and thicker oil in wanner weather.

The chain should also be inspected to ensure that contaminants aren't clinging to it. If an abrasive material is on the chain, it will accelerate wear between the chain parts and the chain and the sprocket.

If wear is observed on the inside of the chain links, it's a sign that the sprockets are out of alignment. Corrective action should be taken before the chain and sprockets are damaged.

If correct tension isn't kept in the chain drive, wear and shock loading will occur to the chain and the sprockets. As a rule, a 2% deflection at the center of the unsupported span is the correct tension for a roller chain drive. Less than that would cause unnecessary loading on the bearings, any more would cause chain slippage and shock loading on the sprocket.

Inspection 4: Typical Belt Conveyor

Refer to Figure B-4 for this section.

Electric Motor

Item 1 is the electric motor. The most common inspection item for the motor is for heat. The motor should be approximately 25 to 29°F high-

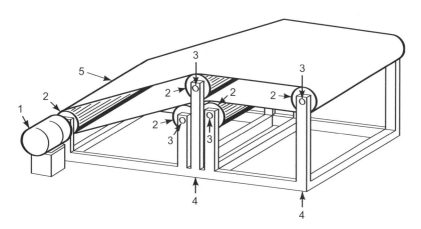

Figure B-4. For inspection 4: typical belt conveyor.

Preventive Maintenance Inspection
Inspection 4: Belt Conveyor

Check the column that indicates the condition of the unit or what problem exists named in the left column.	O.K.	Requires lubrication	Requires adjustment	Requires replacement	Requires cleaning	Excessive vibration	Excessive heat	Loose	See additional comments
1. Electric Motor									
A. Bearings									
B. Base and bolts									
C. Temperature									
D. Vibration									
E. Noise									
2. Coupling Roll									
A. Surface of roll									
B. Alignment									
C. Contamination									
3. Bearing									
A. Free rotation									
B. Excessive play									
C. Visible damage									
D. Lubrication									
E. Heat									
F. Noise									
G. Vibration									
4. Framework									
A. Loose bolts									
B. Broken welds									
C. Bent or broken supports									
5. Belt									
A. Splice condition									
B. Contamination									
C. Tracking									
D. Tension									

Additional Comments:

er in temperature than the surrounding environment. If the motor's temperature is higher than this, it will have a shorter life dueto the deteriorating effect the heat has on the insulation of the motor. Every 20°-temperature rise above the environmental temperature cuts the life of the insulation by one half. When the insulation fails, so does the motor. If the temperature rises above this level, efforts should be made to find the problem so the motor can be cooled back down again.

Another inspection point on the motor is the bearings. The bearings support the rotating part of the motor (usually called the armature or the rotor). The inspection of the bearings should include three basic items: heat, noise, and vibration. These three indicators of trouble may appear singly or in any 'combination. Monitoring these three conditions may require nothing more than the use of the inspector's natural senses. If the equipment is critical in the manufacturing process, it may be advisable to use some form of monitoring or nondestructive testing equipment. For monitoring the temperature, some form of hand-held thermometer may be sufficient. If the equipment is in a difficult-to-reach or unsafe area, some form of temperature-monitoring device may be used. Vibration may be measured by a hand-held meter or a permanent monitoring device. Either of .these types of vibration meters will be able to determine if conditions such as insufficient lubrication, misalignment, contamination, or normal wear are occurring in the bearing. If any of these conditions are detected, then steps should be taken to correct the problem. If they're not corrected, the bearing will fail, and the replacement will also fail quickly.

Another item on the motor to inspect is the base. The base bolts should always be checked to ensure that they're tightened to the correct torque. specifications. If they're not, they may come loose during operation, allowing the motor to misalign with the remaining parts of the drive. Periodically, the base of the motor and the mounting for the motor should be inspected. The foundation may develop faults, allowing the motor to loosen on the base. Settling of the base may occur, allowing it to place excessive stress on the motor base. The base should always be completely clean when installing the motor.

Conveyor Roll

Item 2 is the conveyor roll. This is the roll that the conveyor belt is driven by and that it rides on. The main inspection point for these rolls is the surface. It must be in good condition - free of any sharp edges or defects that could cut the conveyor belt. Keep the rolls free of any contaminants that could damage the roll and/or the belt. If contamination buildup is a problem, steps should be taken to remove the contaminants, and methods should be established to keep the contaminants out of this area. The alignment of the rolls is also important. If the rolls aren't kept in alignment with the conveyor framework and tracking system, excessive wear will occur to the roll and to the belt.

Proper tension of the conveyor is a must because slippage of the drive roll causes heat and wear to develop on the roll and the belt. There are many different tensioning devices in a conveyor system. Consult the manufacturer of each conveyor for recommendations on the proper tension.

Bearings

Item 3 is the bearing that supports the conveyor roll. The bearing should allow free rotation of the roll. If the bearing becomes defective and doesn't allow free rotation of the roll, belt wear and wear of the roll occurs. If allowed to continue, the roll may have a flat spot worn on it by the continual rubbing of the belt. Indications of the bearing condition are heat, noise, and vibration of the hearing. If anyone of these three conditions is excessive, the bearing is showing signs of wear. Consideration then should be given to the degree of wear and whether this necessitates replacement.

Lubrication of the bearing is also important. Most conveyors have a centralized lubrication system, which is supposed to dispense the correct amount of lubricant to each bearing. If the bearing is continually being replaced, the lubrication system should be checked to ensure that good lubricant flow is possible. The line can become plugged up, restricting the amount of lubricant that the bearing receives.

Framework of the Conveyor

Item 4 is the framework of the conveyor. It should be periodically inspected for broken welds or loose bolts. If weld inspections are made, it's good to use a nondestructive testing method, such as magnetic particle, to assist in finding defective welds. The framework must be kept in good shape. If it isn't, the frame may give or twist enough to misalign the rolls, preventing the belt from tracking correctly.

Conveyor Belt

Item 5 is the conveyor belt. The belt carries the load from one location to another. It should be kept in good condition if it's to work properly. Cuts or tears in the belts should be repaired as soon as possible to prevent additional damage to the belt. There are many different types of belts and belting materials. When repairing or splicing a belt, you should consult the manufacturer's recommendation for methods of repairing the belt.

Contamination should be kept off of the belt as much as possible, but especially on the bottom of the belt where it contacts the pulleys. If contamination is allowed in this area, it results in rapid wear of the belt and pulley.

Belt tracking is important to keep the belt centered on the pulleys. If it's allowed to run to one side or another, the belt will be damaged by the conveyor framework or support.

Inspection 5: Typical Chain Conveyor

Refer to Figure B-5 for this section.

Electric Motor

Item 1 is the electric motor. The most common inspection item for the motor is for heat. The motor should be approximately 25 to 29°F higher in temperature than the surrounding environment. If the motor's temperature is higher than this, it will have a shorter life due to the deteriorating effect heat has on the insulation of the motor. For every 20°-temperature rise above the environmental temperature, the life of the insulation is cut by one half. When the insulation fails, so does the motor. If the temperature rises above this level, efforts should be made to find the problem, so the motor can be cooled back down again.

Another inspection point on the motor is bear. The bearings support the rotating part of the motor (usually called the armature or the rotor). The inspection of the bearings should include three basic items: heat, noise, and vibration. These three indicators of trouble may appear singly or in any combination. Monitoring these three conditions may require nothing more than the use of the inspector's natural senses. If the equipment is critical in the manufacturing process, it may be advisable to use some form of monitoring or nondestructive testing equipment. For monitoring the temperature, some form of hand-held thermometer may be sufficient. If the equipment is in a difficult-to-reach or unsafe area, some form of temperature-monitoring device may be used. Vibration may be measured by a hand-held meter or a permanent monitoring device. Either of these types of vibration meters will be able to determine if conditions such as insufficient lubrication, misalignment, contamination, or normal wear are occurring in the bearing. If any of these conditions are detected, then steps should be taken to correct the problem. If they're not corrected, the bearing will fail, and the replacement will also fail quickly.

Another item on the motor to inspect is the base. The base bolts should always be checked to ensure that they're tightened to the correct torque specifications. If they're not, they may come loose during operation, allowing the motor to misalign with the remaining parts of the drive. Periodically, the base of the motor and the mounting for the motor should be inspected. The foundation may develop faults, allowing the motor to loosen on the base. Also, settling of the base may occur, allowing it to place excessive stress on the motor base. The base should always be completely clean when installing the motor.

Coupling

Item 2 in Figure B-5 is the coupling, which is used to connect two shafts together. The most critical item is the alignment of the coupling halves. If they're not aligned within very close tolerances, rapid wear will occur. A rule of thumb is that they must be within 0.005 of an inch. Rigid couplings must be in exact alignment. Flexible couplings may be able to withstand 0.005. The closer the coupling alignment is to being exact, the longer the coupling will last. Misalignment also damages related items in the drive, such as the bearings and shafts. Correct alignment of couplings cannot be overemphasized.

Lubrication is another prune consideration in coupling inspections. If correct lubrication isn't provided for flexible couplings, rapid wear and complete failure will result. If metal-to-metal contact occurs in the coupling, welding and tearing of the coupling material occurs. It's also important not to overlubricate. If overlubrication occurs, fluid friction occurs, which builds up heat and destroys the lubricant, resulting in rapid wear of the coupling material and failure of the coupling. If a coupling is opened up for inspection and a reddish brown color is observed, it should be cleaned and inspected and then properly relubricated.

Conveyor Drive Sprocket

Item 3 is the conveyor. drive sprocket, used to drive the conveyor. The chain must engage and disengage a sprocket in at least two locations.

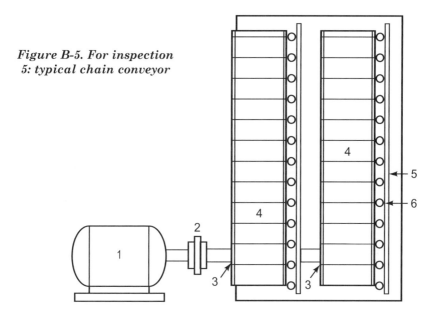

Figure B-5. For inspection 5: typical chain conveyor

The sprocket should be kept in good condition to prevent the chain from hanging in the sprocket and causing unnecessarywear. The sprocket and chain should be inspected to be sure engagement and disengagement is trouble-free.

Alignment is another main inspection point. If the sprockets are misaligned, wear will occur on the sides of the sprockets and inside the links of the chain. These areas need close attention.

Lubrication is also important between the sprocket and the chain. This will prevent excessive wear between the two components.

Conveyor Chain

Item 4 is the conveyor chain. Conveyor chains come in many different types, styles, and sizes. It's best to consult the manufacturer for proper maintenance procedures, but the inspections can be classed in several broad areas. The first would be lubrication. Lubrication is important to give the chain the flexibility to move around the sprocket without damaging the chain. The lubricant must be able to penetrate between all moving parts of the chain. A second consideration is the alignment of the sprockets. Misalignment causes rapid wear on the inside parts of the chain. If this type of wear occurs, the alignment should be checked.

Preventive Maintenance Inspection
Inspection 5: Typical Chain Conveyor

Check the column that indicates the condition of the unit or what problem exists named in the left column.	O.K.	Requires lubrication	Requires adjustment	Requires replacement	Requires cleaning	Excessive vibration	Excessive heat	Loose	See additional comments
1. Electric Motor									
A. Bearings									
B. Base and bolts									
C. Temperature									
D. Vibration									
E. Noise									
2. Coupling									
A. Alignment									
B. Lubrication									
3. Sprocket									
A. Teeth wear									
B. Lubrication									
C. Alignment									
D. Mounting to shaft									
4. Conveyor Chain									
A. Lubrication									
B. Wear									
5. Conveyor Wheels									
A. Lubrication									
6. Track and Frame									
A. Alignment									
B. Loose bolts									
C. Broken welds									
D. Bent or broken supports									

Additional Comments:

The chain should also be examined to ensure that it isn't rubbing or dragging on any part of the supporting framework or floor coverings. This causes wear on both components and could result in enough damage to cause an operational shutdown.

Conveyor Wheels and Tracks
Items 5 and 6 are conveyor wheels and tracks. While these aren't on all conveyors, they're common in heavier industrial applications. This arrangement gives additional support to the conveyor for heavy loads. Lubrication is important to keep the wheels turning on the conveyor. If they won't turn freely, they're of no use. Lubrication is usually administered manually with a power grease gun. It's important that this be carried out on a scheduled basis to prevent wear. The inspector should be alert to the amount of lubricant evidenced on the wheels. The track should be inspected to ensure that it's straight and is supporting the wheels on the conveyor. All framework, bolts, and welds should be inspected to keep the system operating at peak efficiency. Any defect in these areas can lead to an operational shutdown.

Inspection 6: A Typical Hydraulic Lift

Refer to Figure B-6 for this section.

Pump Inlet
Item 1 is the inlet of the pump. The inlet usually has some form of filter on it. The filter should be changed on a regular basis to keep it from becoming stopped up. This will keep it in good condition so it can pass the required amount of fluid on to the pump. This inlet line should always be below the level of the fluid in the tank, thus preventing air from entering the inlet of the pump and damaging the system.

Return Line
Item 2 is the return line from the system. This line carries all the fluid returning to the reservoir. It should enter the tank and stay at a level that will not allow air to enter the returning oil. This will prevent the oil and air mixing and being .carried into the system. The inspector should be alert to keep the oil at a level that will prevent splashing or churning in the tank.

Electric Motor

Item 3 is the electric motor. The most common inspection item for the motor is for heat. The motor should be approximately 25 to 29°F higher in temperature than the surrounding environment. If the motor's temperature is higher than this, it will have a shorter life due to the deteriorating effect the heat has on the insulation of the motor. Every 20°-temperature rise above the environmental temperature cuts the life of the insulation by one half. When the insulation fails, so does the motor. If the temperature rises above this level, efforts should be made to find the problem, so the motor can be cooled back down again.

Another inspection point on the motor is the bearings. The bearings support the rotating part of the motor (usually called the armature or the rotor). The inspection of the bearings should include three basic items: heat, noise, and vibration. These three indicators of trouble may appear singly or in any combination. Monitoring these three conditions may require nothing more than the use of the inspector's natural senses. If the equipment is critical in the manufacturing process, it may be advisable to use some form of monitoring or nondestructive testing equipment. For monitoring the temperature, some form of hand-held thermometer may be sufficient. If the equipment is in a difficult-to-reach or unsafe area, some form of temperature-monitoring device may be used. Vibration. may be measured by a hand-held meter or a permanent monitoring device. Either of these types of vibration meters can determine if conditions such as

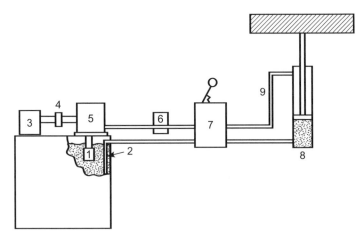

Figure B-6. For inspection 6: simple hydraulic lift.

Preventive Maintenance Inspection
Inspection 6: Hydraulic Lift

Check the column that indicates the condition of the unit or what problem exists named in the left column.	O.K.	Requires lubrication	Requires adjustment	Requires replacement	Requires cleaning	Excessive vibration	Excessive heat	Loose	See additional comments
1. Inlet Filter									
A. Clean									
B. Free intake flow									
2. Return Line									
A. Below fluid level									
3. Electric Motor									
A. Bearings									
B. Base and bolts									
C. Temperature									
D. Vibration									
E. Noise									
4. Coupling									
A. Alignment									
B. Lubrication									
5. Pump									
A. Noise									
B. Flow									
C. Pressure									
D. Base and bolts									
E. Alignment									
F. Leakage									

continued on next page

Inspection 6: continued

Check the column that indicates the condition of the unit or what problem exists named in the left column.	O.K.	Requires lubrication	Requires adjustment	Requires replacement	Requires cleaning	Excessive vibration	Excessive heat	Loose	See additional comments
6. Relief Valve									
A. Adjust pressure									
B. Heat									
7. Directional Control valve									
A. Free Operation									
B. Heat									
8. Hydraulic Cylinder									
A. Leakage									
B. Alignment									
C. Heat									
9. Lines									
A. Mounting secure									
B. Cracks									
C. Loose fittings									
Note: Depending on operating times, the hydraulic fluid should be checked and analyzed to prevent wear on the system.									

Additional Comments:

insufficient lubrication, misalignment, contamination, or normal wear are occurring in the bearing. If any of these conditions are detected, then steps should be taken to correct the problem. If they're not corrected, then the bearing will fail, and the replacement will also fail quickly.

Another item on the motor to inspect is the base. The base bolts should always be checked to ensure that they're tightened to the correct torque specifications. If they're not, they may come loose during operation, allowing the motor to misalign with the remaining parts of the drive, Periodically, the base of the motor and the mounting for the motor should be inspected. The foundation may develop faults, allowing the motor to loosen on the base. Also, settling of the base may occur, causing excessive stress on the motor base. The base should always be completely clean when installing the motor.

Couplings

Item 4 is the coupling, which is used to connect two shafts together. The most critical item is the alignment of the coupling halves. If they're not aligned within very close tolerances, rapid wear will occur. A rule of thumb is that they must be within 0.005 of an inch. Rigid couplings must be in exact alignment. Flexible couplings may be able to withstand 0.005. The closer the coupling alignment is to being exact, the longer the coupling will last. Misalignment also damages related items in the drive, such as the bearings and shafts. Correct alignment of couplings cannot be overemphasized.

Lubrication is another prime consideration in coupling inspections. If correct lubrication isn't provided for flexible couplings, rapid wear and complete failure will result. If metal-to-metal contact occurs in the coupling, welding and tearing of the coupling material occurs. It's also important not to overlubricate. Overlubrication causes fluid friction, which causes heat buildup and destroys the lubricant. This then results in rapid wear of the coupling material and failure of the coupling. If a coupling is opened up for inspection and a reddish brown color is observed, it should be cleaned and inspected and then properly relubricated.

Pump

Item 5 is the pump, which is used to produce the necessary flow in the system. The inspector can usually determine the problem with the pump by listening to it. A hydraulic pump is supposed to run quietly. A growling noise indicates a problem with the system. Cavitation may be occurring or air may be getting into the inlet. A quick check of the reservoir may give an answer. If the oil looks normal and the pump is growling, it's probably cavitation. If the pump is growling and the oil is milky in the reservoir, it's probably air getting into the inlet.

The pump also has base bolts. Make sure the bolts are tight. Any looseness will allow the pump to shift and cause misalignment, which will damage the pump and motor. From time to time, alignment should be checked to ensure that no shifting has taken place. If bearing wear occurs or if wear occurs in the pumping element, the output and efficiency of the system will diminish. If this is suspect,

the inspector may want to use a flow and pressure check to determine if the pump is worn enough to be replaced.

Pressure Relief Valve

Item 6 is the pressure relief valve. This valve is used to control maximum system pressure. It should be set at the level specified by the system manufacturer. This valve should be dismantled from time to time to check for worn or broken parts. If the system pressure drops or is not consistent, this valve may be the problem.

There may be an unloading valve in this area to dump the load of the pump to the reservoir at a lower pressure to prevent overheating of the pressure relief valve. If the relief valve is excessively hot, the pressure setting of the relief and unloading valve should be examined.

Directional Control Valve

Item 7 is the directional control valve. This valve can be operated by various means: electrical, pneumatic, mechanical, or manual. Its purpose is to control the direction of the actuator. One main consideration with this valve is heat. If the temperature is excessive, it indicates a possible internal leak. This will mean that wear has occurred that allows the valve to pass fluid back to the tank. It would be advisable to dismantle the valve for a visual inspection. From time to time, dirt, gums, and varnishes can build up and prevent free movement of the valve. It will be necessary to dismantle the valve to clean it.

Hydraulic Cylinder

Item 8 is the hydraulic cylinder. This is the device being controlled in the system. The cylinder should extend and retract freely with no load. If the cylinder fails to move and the rest of the system is operating correctly, it's possible for the cylinder to be passing fluid through its seals. If this is the case, it will be necessary to dismantle the cylinder and replace the seals. Other inspection points include the rod seal, which should prevent oil from flowing out of the cylinder rod, and the mounting of the cylinder. If the mounting isn't correct, it will allow too much play, causing premature wear of the seals.

Lines

Item 9 indicates the lines. The lines should always be fastened or anchored securely. This prevents movement of the lines during the operation of the system. If they're not anchored, the movement will wear the lines and cause loose fittings or broken welds. The inspector should be alert for any line leakage.

Inspection 7: A Typical Hydraulic Motor Circuit

Refer to Figure B-7 for this section.

Pump Inlet

Item 1 is the inlet of the pump. The inlet usually has some form of filter on it. The filter should be changed on a regular basis to keep it from becoming stopped up. This will keep it in good condition so it can pass the required amount of fluid on to the pump. This inlet line should always be below the level of the fluid in the tank, thus preventing air from entering the inlet of the pump and damaging the system.

Return Line

Item 2 is the return line from the system. This line carries all the fluid returning to the reservoir. It should enter the tank and stay at a level that will not allow air to enter the returning oil. This will prevent the oil and air mixing and being carried into the system. The inspector should be alert to keep the oil at a level that will prevent splashing or churning in the tank.

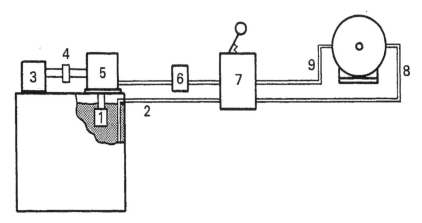

Figure B-7 For inspection 7 :simple hydraulic motor

Electric Motor

Item 3 is the electric motor. The most common inspection item for the motor is for heat. The motor should be approximately 25 to 29°F higher in temperature than the surrounding environment. If the motor's temperature is higher than this, it will have a shorter life due to the deteriorating effect the heat has on the insulation of the motor. Every 20°-temperature rise above the environmental temperature cuts the life of the insulation by one half. When the insulation fails, so does the motor. If the temperature rises above this level, efforts should be made to find the problem, so the motor can be cooled back down again.

Another inspection point on the motor is the bearings. The bearings support the rotating part of the motor (usually called the armature or the rotor). The inspection of the bearings should include three basic items: heat, noise, and vibration. These three indicators of trouble may appear singly or in any combination. Monitoring these three conditions may require nothing more than the use of the inspector's natural senses. If the equipment is critical in the manufacturing process, it may be advisable to use some form of monitoring or nondestructive testing equipment. For monitoring the temperature, some form of hand-held thermometer may be sufficient. If the temperature is in a difficult-to-reach or unsafe area, some form of temperature-monitoring device may be used. Vibration may be measured by a hand-held meter or a permanent monitoring device. Either of these types of vibration meters will be able to determine if conditions such as insufficient lubrication, misalignment, contamination, or normal wear are occurring in the bearing. If any of these conditions are detected, then steps should be taken to correct the problem. If they're not corrected, then the bearing will fail, and the replacement. will also fail quickly.

Another item on the motor to inspect is the base. The base bolts should always be checked to ensure that they're tightened to the correct torque specifications. If they're not, they may come loose during operation, allowing the motor to misalign with the remaining parts of the drive. Periodically, the base of the motor and the mounting for the motor should be inspected. It's possible that the foundation may develop faults, allowing the motor to loosen on the base. Also, settling of the base may occur, causing excessive stress on the motor base. The base should always be completely clean when installing the motor.

Couplings

Item 4 is the coupling, which is used to connect two shafts together. The most critical item is the alignment of the coupling halves. If they're not aligned within very close tolerances, rapid wear will occur. A rule of thumb is that they must be within 0.005 of an inch. Rigid couplings must be in exact alignment. Flexible couplings may be able to withstand 0.005. The closer the coupling alignment is to being exact, the longer the coupling will last. Misalignment also damages related items in the drive, such as the bearings and shafts. Correct alignment of couplings cannot be overemphasized.

Lubrication is another prime consideration in coupling inspections. If correct lubrication isn't provided for flexible couplings, rapid wear and complete failure will result. If metal-to-metal contact occurs in the coupling, welding and tearing of the coupling material occurs. It's also important not to overlubricate. Overlubrication causes fluid friction, which builds up heat and destroys the lubricant, resulting in rapid wear of the coupling material and failure of the coupling. If a coupling is opened up for inspection and a reddish brown color is observed, it should be cleaned and inspected and then properly relubricated.

Pump

Item 5 is the pump. The pump is used to produce the necessary flow in the system. The inspector can usually determine the problem with the pump by listening to it. A hydraulic pump is supposed to run quietly. A growling noise indicates a problem with the system. Cavitation may be occurring or air may be getting into the inlet. A quick check of the reservoir may give an answer. If the oil looks normal and the pump is growling, it's probably cavitation. If the pump is growling and the oil is milky in the reservoir, it's probably air getting into the inlet. The pump also has base bolts, which should be tight. Any looseness will allow the pump to shift and cause misalignment, which damages the pump and motor. From time to time, alignment should be checked to ensure that no shifting has taken place. If bearing wear occurs or if wear occurs in the pumping element, the output and efficiency of the system will diminish. If this is suspect, the inspector may want to use a flow and pressure check to determine if the pump is worn enough to be replaced.

Pressure Relief Valve

Item 6 is the pressure relief valve, which is used to control maximum system pressure. It should be set at the level specified by the system manufacturer. This valve should be dismantled from time to time to check for worn or broken parts. If the system pressure drops or isn't consistent, this valve may be the problem.

Also, there may be an unloading valve in this area to dump the load of the pump to the reservoir at a lower pressure to prevent overheating of the pressure relief valve. If the relief valve is excessively hot, the pressure setting of the relief and unloading valve should be examined.

Directional Control Valve

Item 7 is the directional control valve. This valve can be operated by various means: electrical, pneumatic, mechanical, or manual. Its purpose is to control the direction of the actuator. One main consideration with this valve is heat. Excessive temperature indicates a possible internal leak. Such a condition means that wear has occurred which allows the valve to pass fluid back to the tank. It's advisable to dismantle the valve for a visual inspection. Dirt, gums, and varnishes build up from time to time and prevent free movement of the valve. In this case, it's necessary to dismantle the valve to clean it.

Hydraulic Motor

Item 8 in Figure B-7 is the hydraulic motor. This is the device being controlled in the system. The motor should rotate freely with no load. If the motor fails to move and the rest of the system is operating correctly, it's possible that the motor is passing fluid through its seals. If this is the case, it will be necessary to dismantle the motor and replace the seals. Another possibility is that the internal components have worn to a degree that allows free passage of fluid. If this is the case, replacement of the components or the motor will be required. Other inspection points include the shaft seal, which should prevent oil from flowing out of the motor shaft, and the mounting ofthe motor. Incorrect mounting allows too much play and causes premature wear of the seals.

Lines

Item 9 denotes the lines. The lines should always be fastened or anchored securely to prevent their movement during operation of the sys-

Preventive Maintenance Inspection
Inspection 7: Hydraulic Motor

Check the column that indicates the condition of the unit or what problem exists named in the left column.	O.K.	Requires lubrication	Requires adjustment	Requires replacement	Requires cleaning	Excessive vibration	Excessive heat	Loose	See additional comments
1. Inlet Filter									
A. Clean									
B. Free intake flow									
2. Return Line									
A. Below fluid level									
3. Electric Motor									
A. Bearings									
B. Base and bolts									
C. Temperature									
D. Vibration									
E. Noise									
4. Coupling									
A. Alignment									
B. Lubrication									
5. Pump									
A. Noise									
B. Flow									
C. Pressure									
D. Base and bolts									
E. Alignment									
F. Leakage									

continued on next page

Inspection 7: continued

Check the column that indicates the condition of the unit or what problem exists named in the left column.	O.K.	Requires lubrication	Requires adjustment	Requires replacement	Requires cleaning	Excessive vibration	Excessive heat	Loose	See additional comments
6. Relief Valve									
A. Adjust pressure									
B. Heat									
7. Directional Control valve									
A. Free Operation									
B. Heat									
8. Hydraulic Cylinder									
A. Leakage									
B. Alignment									
C. Heat									
9. Lines									
A. Mounting secure									
B. Cracks									
C. Loose fittings									
Note: Depending on operating times, the hydraulic fluid should be checked and analyzed to prevent wear on the system.									

Additional Comments:

tem. If they're not anchored, the movement will wear the lines and cause loose fittings or broken welds. The inspector should be alert for any line leakage.

Inspection 8: Typical Pneumatic Cylinder
Refer to Figure B-8 for this section.

Electric Motor
Item 1 in the figure is the electric motor. The most common in-spection item for the motor is for heat. The motor should be approx-imately 25 to 29°F higher in temperature than the surrounding environ-ment. If the motor's temperature is higher than this, it will have a shorter life due to the deteriorating effect the heat has on the insulation of the motor. Every 20°-temperature rise above the environmental temperature cuts the life of the insulation by one half. When the insulation fails, so does the motor. If the temperature rises above this level, efforts should be made to find the problem, so the motor can be cooled back down again.

Another inspection point on the motor is the bearings, which sup-port the rotating part of the motor (usually called the armature or the rotor). The inspection of the bearings should include three basic items: heat, noise, and vibration. These three indicators of trouble may appear singly or in any combination. Monitoring these three conditions may require nothing more than the use of the inspector's natural senses. If the equipment is critical in the manufacturing process, it may be advisable to use some form of monitoring or nondestructive testing equipment. For monitoring the temperature, some form of hand-held thermometer may be sufficient. If the equipment is in a difficult-to-reach or unsafe area, some form of temperaturemonitoring device may be used. Vibration may be measured by a hand-held meter or a permanent monitoring device. Either of these types of vibration meters can determine if conditions such as

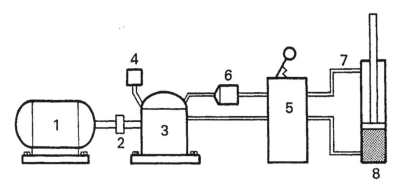

Figure B-8. For inspection 8: simple pneumatic cylinder.

Preventive Maintenance Inspection
Inspection 8: Pneumatic Cylinder

Check the column that indicates the condition of the unit or what problem exists named in the left column.	O.K.	Requires lubrication	Requires adjustment	Requires replacement	Requires cleaning	Excessive vibration	Excessive heat	Loose	See additional comments
1. Electric Motor									
A. Bearings									
B. Base and bolts									
C. Temperature									
D. Vibration									
E. Noise									
2. Coupling									
A. Alignment									
B. Lubrication									
3. Compressor									
A. Flow									
B. Pressure									
C. Noise									
D. Vibration									
E. Lubrication									
F. Heat									
4. Inlet Filter									
A. Clean									
B. Free flow									
5. Directional Control Valve									
A. Free movement									
B. Proper air flow									

continued on next page

Inspection 8: continued

Check the column that indicates the condition of the unit or what problem exists named in the left column.	O.K.	Requires lubrication	Requires adjustment	Requires replacement	Requires cleaning	Excessive vibration	Excessive heat	Loose	See additional comments
6. Muffler									
A. Quiet									
B. Good flow									
7. Lines									
A. Properly mounted									
B. Leaks									
C. Loose fittings									
D. Broken piping									
8. Pneumatic Cylinder									
A. Free movement									
B. Good alignment									
C. Proper mounting									
D. Leakage									

Additional Comments:

insufficient lubrication, misalignment, contamination, or normal wear are occurring in the bearing. If any of these conditions are detected, then steps should be taken to correct the problem. If they're not corrected, the bearing will fail, and the replacement will also fail quickly.

Another item on the motor to inspect is the base. The base bolts should always be checked to ensure that they're tightened to the correct torque specifications. If they're not, they may come loose during operation, allowing the motor to misalign with the remaining parts of the drive. Periodically, the base of the motor and the mounting for the motor should be inspected. The foundation may develop faults, allowing the motor to loosen on the base; or settling of the base may occur, causing excessive stress on the motor base. The base should always be completely clean when installing the motor.

Couplings

Item 2 is the coupling, which is used to connect two shafts together. The most critical item is the alignment of the coupling halves. If they're not aligned within very close tolerances, rapid wear will occur. A rule of thumb is that they must be within 0.005 of an inch. Rigid couplings must be in exact alignment. Flexible couplings may be able to withstand 0.005. The closer the coupling alignment is to being exact, the longer the coupling will last. Misalignment also damages related items in the drive, such as the bearings and shafts. Correct alignment of couplings cannot be overemphasized.

Lubrication is another prime consideration in coupling inspections. If correct lubrication isn't provided for flexible couplings, rapid wear and complete failure will result. If metal-to-metal contact occurs in the coupling, welding and tearing of the coupling material occurs. Ws also important not to overlubricate. Overlubrication causes fluid friction, which builds up heat and destroys the lubricant. In this case, rapid wear of the coupling material and failure of the coupling will occur. If a coupling is opened up for inspection and a reddish brown color is observed, it should be cleaned and inspected and then properly relubricated.

Compressor

Item 3 is the compressor, which changes the mechanical energy into pneumatic energy Most compressors have a pumping chamber equipped with valves and seals to properly compress the air. If either of the valves (inlet or discharge) or the seals are leaking, the output of the pump won't be as high as it should be. If the compressor cannot keep up with system demand, but is rated high enough, the inspector should recommend dismantling and rebuilding the compressor. Most compressors have a crankcase or gearcase for the storage of lubricant. The lubricant should always be kept at the proper level to prevent damage to the mechanical components in the compressor.

Some compressors are equipped with internal coolers to lower the temperature of the air. Coolers may be air or water cooled. If excessive moisture is in the air after compression, the cooler should be checked for leakage. If the air is still hot after being run through the cooler, the inspector should check the cooler to see if it's stopped up and not allowing good flow of the cooling medium.

Inlet Filter

Item 4 is the inlet filter. All air in the system must at one time come through this filter. The inlet should always be positioned in a clean, dry location for efficient operation of the system. The filter should be periodically changed, otherwise the compressor can't draw enough air through the filter, and the system will operate at a high temperature and in a sluggish manner. Any contamination that passes through this filter is sure to cause problems in the system.

Directional Control Valve

Item 5 is the directional control valve. This component controls the direction of the actuator and may be activated by mechanical, pneumatic, or electrical means. The valve has a spool inside that shifts, directing the air flow to an outlet port. The port out of which the air flow is directed determines the direction of the device being controlled. If the directional control valve doesn't shift freely, it should be dismantled and inspected. The valve should be cleaned and all defective parts replaced before it is reassembled.

Muffler

Item 6 is the muffler. It's used to quiet the exhaust of the system as it's returned to the atmosphere. There are usually two main types of mufflers. One is freeflow and the other is adjustable. The adjustable one can be used to control the flow rate of the air as it's returned to the atmosphere. This will allow for speed control of the. actuator in the circuit. If either type becomes clogged or restricted, it will affect operation of the circuit. The muffler should always be inspected for proper operation.

Lines Carrying Compressed Air

Item 7 shows the lines carrying the compressed air. The lines and related fittings should always be checked for leakage. If the lines are allowed to leak, the compressor won't be able to supply enough air to operate the system properly. There are some listening devices on the market that will detect air leaks at considerable distances. If the lines are in locations that are difficult to reach, these devices are very convenient. The lines should also be checked for rigidity. They should be clamped and held in place to prevent motion and vibration, both of which will cause eventual loosening or cracking of the lines.

Pneumatic Cylinder

Item 8 is the pneumatic cylinder. Pneumatic cylinders are usually used for high-speed low-power applications. Inspection points include proper motion, looseness of mounting devices, leakage of any fittings, and leakage at the rod of the cylinder. Any of these defects need attention as soon as it's possible to schedule the repair.

Inspection 9: Typical Pneumatic Motor

Refer to Figure B-9 for this section.

Electric Motor

Item 1 is the electric motor. The most common inspection item for the motor is for heat. The motor should be approximately 25 to 29°F higher in temperature than the surrounding environment. If the motor's temperature is higher than this, it will have a shorter life due to the deteriorating effect the heat has on the insulation of the motor. Every 20°-temperature rise above the environmental temperature cuts the life of the insulation by one half. When the insulation fails, so does the motor. If the temperature rises above this level, efforts should be made to find the problem, so the motor can be cooled back down again.

Another inspection point on. the motor is the bearings. The bearings support the rotating part of the motor (usually called the armature or the rotor). The inspection of the bearings should include three basic items: heat, noise, and vibration. These three indicators of trouble may appear singly or in any combination. Monitoring these three conditions may require nothing more than the use of the inspector's natural senses. If the equipment is critical in the manufacturing process, it may be advisable to use some form of monitoring or nondestructive testing equipment. For

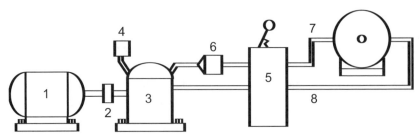

Figure B-9. For inspection 9: simple pneumatic motor.

Preventive Maintenance Inspection
Inspection 9: Pneumatic Motor

Check the column that indicates the condition of the unit or what problem exists named in the left column.	O.K.	Requires lubrication	Requires adjustment	Requires replacement	Requires cleaning	Excessive vibration	Excessive heat	Loose	See additional comments
1. Electric Motor									
A. Bearings									
B. Base and bolts									
C. Temperature									
D. Vibration									
E. Noise									
2. Coupling									
A. Alignment									
B. Lubrication									
3. Compressor									
A. Flow									
B. Pressure									
C. Noise									
D. Vibration									
E. Lubrication									
F. Heat									
4. Inlet Filter									
A. Clean									
B. Free flow									
5. Directional Control Valve									
A. Free movement									
B. Proper air flow									

continued on next page

Inspection 9: continued

Check the column that indicates the condition of the unit or what problem exists named in the left column.	O.K.	Requires lubrication	Requires adjustment	Requires replacement	Requires cleaning	Excessive vibration	Excessive heat	Loose	See additional comments
6. Muffler									
A. Quiet									
B. Good flow									
7. Lines									
A. Properly mounted									
B. Leaks									
C. Loose fittings									
D. Broken piping									
8. Pneumatic Cylinder									
A. Free movement									
B. Good alignment									
C. Proper mounting									
D. Leakage									

Additional Comments:

monitoring the temperature, some form of hand-held thermometer may be sufficient. If the equipment is in a difficult-to-reach or unsafe area, some form of temperature-monitoring device may be used. Vibration may be measured by a hand-held meter or a permanent monitoring device. Either of these types of vibration meters can determine if conditions such as insufficient lubrication, misalignment, contamination, or normal wear are occurring in the bearing. If any of these conditions are detected, then steps should be taken to correct the problem. If they're not corrected, the bearing will fail, and the replacement will also fail quickly.

Another item on the motor to inspect is the base. The base bolts should always be checked to ensure that they're tightened to the correct torque specifications. If they're not, they may come loose during operation, allowing the motor to misalign with the remaining parts of the drive.

Periodically, the base of the motor and the mounting for the motor should be inspected. The foundation may develop faults, allowing the motor to loosen on the base. Also, settling of the base may occur, causing excessive stress on the motor base. The base should always be completely clean when installing. the motor.

Couplings

Item 2 is the coupling, which is used to connect two shafts together. The most critical item is the alignment of the coupling halves. If they're not aligned within very close tolerances, rapid wear will occur. A rule of thumb is that they must be within 0.005 of an inch. Rigid couplings must be in exact alignment. Flexible couplings may be able to withstand 0.005. The closer the coupling alignment is to being exact, the longer the coupling will last. Misalignment, however, can damage related items in the drive, such as the bearings and shafts. Correct alignment of couplings cannot be overemphasized.

Lubrication is another prime consideration in coupling inspections. If correct lubrication isn't provided for flexible couplings, rapid wear and complete failure will result. If metal-to-metal contact occurs in the coupling, welding and tearing of the coupling material occurs. Along this same line, it's also important not to overlubricate. Overlubrication causes fluid friction, which builds up heat and destroys the lubricant. This condition results in rapid wear of the coupling material and failure of the coupling. If a coupling is opened up for inspection and a reddish brown color is observed, it should be cleaned and inspected and then properly relubricated.

Compressor

Item 3 is the compressor, the device that changes the mechanical energy into pneumatic energy. Most compressors have a pumping chamber equipped with valves and seals to properly compress the air. If either of the valves (inlet or discharge) or the seals are leaking, the output of the pump won't be as high as it should. If the compressor cannot keep up with system demand, but is rated high enough, the inspector should: recommend dismantling and rebuilding the compressor. Most compressors have a crankcase or gearcase for the storage of lubricant. The lubricant should always be kept at the proper level to prevent damage to the mechanical components in the compressor. Some compressors are equipped with internal coolers to lower the temperature of the air. Coolers may be air or water cooled. If excessive moisture is in the air after compression, the cooler should be checked for leakage. If the air is still hot after being run through the cooler, the inspector should check the cooler to see if it's stopped up and not allowing good flow of the cooling medium.

Inlet Filter

Item 4 is the inlet filter. This filter is important because all air in the system must at one time come through this filter. The inlet should always be positioned in a clean, dry location for efficient operation of the system. The filter should be periodically changed, otherwise the compressor can't draw enough air through the filter, and the system will operate at a high temperature and in a sluggish manner. Any contamination that passes through this filter is sure to cause problems in the system.

Directional Control Valve

Item 5 is the directional control valve. This component controls the direction of the actuator and may be activated by mechanical, pneumatic, or electrical means. The valve has a spool inside that shifts, directing the air flow to an outlet port. The port out of which the air flow is directed determines the direction of the device being controlled. If the directional control valve doesn't shift freely, it should be dismantled and inspected. The valve should be cleaned and all defective parts replaced before it is reassembled.

Muffler

Item 6 is the muffler, which is used to quiet the exhaust of the system as it's returned to the atmosphere. There are usually two main types of mufflers. One is free flow and the other is adjustable. The adjustable one can be used to control the flow rate of the air as it's returned to the atmosphere. This will allow for speed control of the actuator in the circuit. If either type becomes clogged or restricted, it will affect operation of the circuit. The muffler should always be inspected for proper operations.

Lines Carrying Compressed Air

Item 7 shows the lines carrying the compressed air. The lines and related fittings should always be checked for leakage. If the lines are allowed to leak, the compressor won't be able to supply enough air to operate the system properly. There are some listening devices on the market that detect air leaks at considerable distances. If the lines are in locations that are difficult to reach, these devices are very convenient. The lines should also be checked for rigidity. They should be clamped and held in place to prevent motion and vibration, both of which will cause eventual loosening or cracking of the lines.

Pneumatic Motor

Item 8 is the pneumatic motor. Pneumatic motors are usually used for high-speed low-torque applications. Inspection points indude proper motion, looseness of mounting devices, leakage of any fittings, and leakage at the rod of the cylinder. Any of these defects need attention as soon as it's possible to schedule the repair.

Inspection 10: A Simple Electrical Starter

Refer to Figure B-10 for this section.

Incoming Wiring

Item 1 is the incoming wiring. This wiring should be inspected for worn or defective insulation. Any points where the wires make sharp bends or touch any metal supports are good places to begin inspecting. Frayed insulation or exposed wiring is a condition needing immediate attention. The wires should also be inspected where they fasten to the starter, whether this is on a terminal block, or if they fasten directly to the relay. The wires should be checked for looseness, because this causes heating and discoloration of the terminal. Discoloration means that the fitting is loose or has been loose, and thus should be closely checked for proper tightness.

Terminals in the Starter Panel

Item 2 represents the terminals in the starter panel. The main inspection point of the terminals is tightness. If they work loose, then high current develops at this point. High current develops high heat, which may be intense enough to weld the terminal. If this occurs, replacement of the terminal is required. Any terminal that has been overheated will be discolored, which is a good visible sign to the inspector.

Control Transformer

Item 3 is the control transformer, a device used to lower the line voltage. The two most common industrial sizes are 220 to 110 V and 440 to 220 V. The terminals and the insulation of the transformer are the two main inspection points. Correct tightness of the terminals is important. The insulation on the transformer should be visually inspected to ensure that it's not breaking or peeling off the transformer. This condition exposes the wiring and presents a possible hazard or malfunction. All insulation should be kept in good condition.

Contact Tips

Item 4 shows the contact tips in the relay. In some cases, it's necessary to remove a shield to inspect these tips, which are usually made of

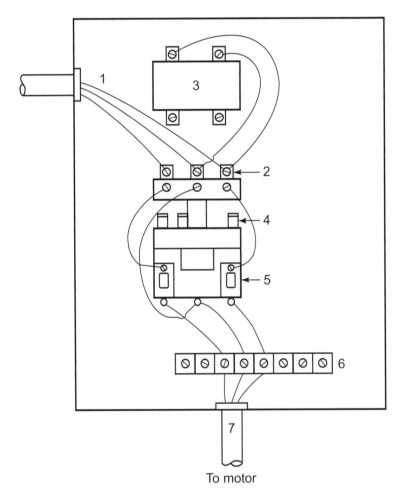

To motor

Figure B-10. For inspection 10: simple electrical starter.

Preventive Maintenance Inspection
Inspection 10: Electrical Starter

Check the column that indicates the condition of the unit or what problem exists named in the left column.	O.K.	Requires lubrication	Requires adjustment	Requires replacement	Requires cleaning	Excessive vibration	Excessive heat	Loose	See additional comments
1. Incoming Wiring									
A. Insulation									
B. Loose connections									
C. Heat									
2. Terminals									
A. Proper mountings									
B. Loose connections									
C. Heat									
3. Transformer									
A. Proper mounting									
B. Insulation									
C. Loose connections									
D. Heat									
4. Relay Tips									
A. Wear									
B. Springs									
C. Heat									
5. Overload Relay									
A. Proper size									
B. Tip condition									
C. Loose connections									
D. Heat									
8. Terminal Strip									
A. Correct mounting									
B. Loose connections									
C. Loose									
7. Outside Leads									
A. Insulation									
B. Loose connections									
C. Heat									

Additional Comments:

a silver-coated copper alloy. If they're worn sufficiently, replacement is necessary to ensure proper operation of the relay. Springs are used to compensate for wear of the tips. These should be in good condition and in their proper location, otherwise the relay won't operate correctly, and burning of the tips will result. If the tips are sticking or welding together, the spring tension should be checked. If the tips stick together periodically, replacement should be considered.

Overload Coil and Tips

Item 5 is the overload coil and tips. The line current is passed through these tips. If the current becomes excessive, the heat will open these tips and drop the relay out, preventing overloads. The problem occurs when the overload is changed. If too large an overload is put in, it will damage the system and never drop the relay out. If too small an overload is installed, it will open the circuit too often, causing more service calls than necessary. The inspector should check to be sure that the correct-sized overload is in the relay. Tightness of all connections is a must if the current through the overload is to be a proper indication of current in the circuit.

Terminal Strip

Item 6 is the terminal strip, a device that connects the relay panel to the wiring in the rest of the circuit. Tightness of the connections on the strip is very important. Also, the wires should be spaced far enough apart so that as they go to different terminals, they don't make contact. Neat, well-arranged wiring is very helpful to the inspector and repairman.

Wiring and Related Hardware

Item 7 is the wiring and related hardware going to a motor. The wiring should be in good condition, with good insulation. The conduit or flexible covering for the wires should be firmly anchored to prevent unnecessary motion or vibration. This will prevent additional wear on the wires. All devices should be securely mounted in the panel. If the components are allowed to move, they may short out against each other or flex the wire enough to break it.

Inspection 11: A Typical Lighting Circuit

Refer to Figure B-11 for this section.

Circuit Breaker

Item 1 is the circuit breaker that supplies power to the circuit. The main inspection point of this component is the tightness of all connections. Tight connections prevent most problems. If the breaker begins tripping frequently, it's usually a sign of wear in the breaker. If there is no problem in the circuit, consideration may be given to rebuilding or replacing the breaker.

On-Off Switch

Item 2 is the on-off switch for the lights. The main inspection point here is the correct connection and protection of the wiring. If the wires are connected and tightened correctly, most problems will be minimal. The wiring should be in conduit or some form of flexible covering to prevent damage to the wires. If they're not protected, any small breakdown in insulation will cause a lighting failure.

Relay Panel

Item 3 is the relay panel. This item is optional, although it is found in heavier industrial lighting. The relay panel should be given the same basic inspection as the relay panel in the previous example. Almost all components will be similar.

Ballast

Item 4 is a ballast of the transformer. This device is used to change the line voltage to the correct level for the light. Proper connection of the

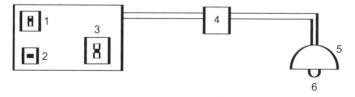

Figure B-11. For inspection 11: typical lighting circuit.

Preventive Maintenance Inspection
Inspection 11: Typical Lighting Circuit

Check the column that indicates the condition of the unit or what problem exists named in the left column.	O.K.	Requires lubrication	Requires adjustment	Requires replacement	Requires cleaning	Excessive vibration	Excessive heat	Loose	See additional comments
1. Incoming Wiring									
A. Wire insulation									
B. Loose terminals									
C. Heat									
2. Off-On Switch									
A. Wire insulation									
B. Loose terminals									
C. Heat									
3. Relay Panel									
A. See Previous Inspection (#10)									
4. Ballast									
A. Wire insulation									
B. Loose terminals									
C. Heat									
5. Light Shade and Socket									
A. Cleanliness									
B. Loose connections									
C. Insulation									
D. Heat									
6. Light Bulb									
A. Loose connections									
B. Insulation									
C. Correct wattage									

Additional Comments:

wiring and tightness of all fittings is the primary inspection.

Light Shade and Socket

Item 5 represents the light shade and socket. The shade may require cleaning from time to time, especially in dirty environments. The socket requires occasional inspection (usually during the lamp replacement). Loose wires, screws, or connecting components are obvious signs of trouble. The mounting should always be adequate, for if the light vibrates, the filament in the bulb won't last very long.

Light Bulb

Item 6 is the light bulb. It should always be the correct wattage and voltage for the light. A larger wattage can cause an overload on the socket or too large a load on the rest of the system. A light that's too small could result in insufficient lighting and unsafe conditions.

Inspection 12: Typical Motor-Generator Set

Refer to Figure B-12 for this section.

Electric Motor

Item 1 is the electric motor. The most common inspection item for the motor is for heat. The motor should be approximately 25 to 29°F higher in temperature than the surrounding environment. If the motor's temperature is higher than this, it will have a shorter life due to the deteriorating effect the heat has on the insulation of the motor. Every 20°-temperature rise above the environmental temperature cuts the life of the insulation by one half. When the insulation fails, so does the motor. If the temperature rises above this level, efforts should be made to find the problem, so the motor can be cooled back down again.

Another inspection point on the motor is the bearings, which support the rotating part of the motor (usually called the armature or the rotor). The inspection of the bearings should include three basic items: heat, noise, and vibration. These three indicators of trouble may appear singly or in any combination. Monitoring these three conditions may require nothing more than the use of the inspector's natural senses. If the equipment is critical in the manufacturing process, it may be advisable to use some form of monitoring or nondestructive testing equipment. For

Preventive Maintenance Inspection
Inspection 12: Typical Motor-Generator Set

Check the column that indicates the condition of the unit or what problem exists named in the left column.	O.K.	Requires lubrication	Requires adjustment	Requires replacement	Requires cleaning	Excessive vibration	Excessive heat	Loose	See additional comments
1. Electric Motor									
A. Bearings									
B. Base and bolts									
C. Temperature									
D. Vibration									
E. Noise									
2. Coupling									
A. Alignment									
B. Lubrication									
3. Generator									
A. All of electric motor (#1)									
B. Armature									
C. Brushes									
D. Rotor									

Additional Comments:

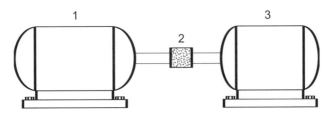

Figure B-12. For inspection 12: motor-generator set.

monitoring the temperature, some form of hand-held thermometer may be sufficient. If the equipment is in an area that's difficult to reach or is unsafe, some form of temperature-monitoring device may be used. Vibration may be measured by a hand-held meter or a permanent monitoring device. Either of these types of vibration meters is able to determine if conditions such as insufficient lubrication, misalignment, contamination, or normal wear are occurring in the bearing. If any of these conditions are detected, then steps should be taken to correct the problem. If they're not corrected, . then the bearing will fail, and the replacement will also fail quickly.

Another item on the motor to inspect is the base. The base bolts should - always be checked to ensure that they're tightened to the correct torque specifications. If they're not, they may come loose during operation, allowing the motor to misalign with the remaining parts of the drive. Periodically, the base of the motor and the mounting for the motor should be inspected. The foundation may develop faults, allowing the motor to loosen on the base. Settling of the base may occur, causing excessive stress on the motor base. The base should always be completely clean when installing the motor.

Couplings

Item 2 is the coupling, which is used to connect two shafts together. The most critical inspection item is the alignment of the coupling halves. If they're not aligned within very close tolerances, rapid wear will occur. A rule of thumb is that they must be within 0.005 of an inch. Rigid couplings must be in exact alignment. Flexible couplings may be able to withstand 0.005. The closer the coupling alignment is to being exact, the longer the coupling will last. Misalignment will also damage related items in the drive, such as the bearings and shafts. Correct alignment of couplings cannot be overemphasized.

Lubrication is another prime consideration in coupling inspections. If correct lubrication isn't provided for flexible couplings, rapid wear and complete failure will result. If metal-to-metal contact occurs in the coupling, welding and tearing of the coupling material occurs. It's important not to overlubricate. Overlubrication causes fluid friction, which creates heat buildup and destroys the lubricant. This condition results in rapid wear of the coupling material and failure of the coupling. If a coupling is opened up for inspection and a reddish brown color is observed, it should be cleaned and inspected and then properly relubricated.

Generator

Item 3 is the generator, which is the output of the system. The usual inspection points include the armature, the rotor, and the brushes. In addition to these areas, the motor inspection points should be considered. A generator is usually nothing more than a motor that is mechanically driven.

APPENDIX C
SAMPLE COMPONENT PM INSPECTIONS

Air Compressor Control Panel

Equipment Data

Production Unit	
Field	
Platform	
Service	Air
System	
Component	Control Panel

PM Data

Frequency	Quarterly
Priority	
Use Procedure	
Procedure file	
Craft	Electrical
Shutdown Req ?	**No**
Total Duration	1 hour
MMS ?	**No**

Before proceeding with this work, the equipment needs to be safe. This means that all potential energy sources, such as pressure or power, need to be positively identified and locked out/tagged out according to your platform specific LO/TO procedures for this of equipment.

Equipment List

SAP Identifier	
SAP Identifier	

Figure C-1 Air Compressor Control Panel

Application		Location	Switchgear Room
Drawing Number			
SAP Identifier			
Spare		Spare	

Validate Specific Equipment Data	OK	NOT OK	List Missing or Incorrect Data
As Found	◯	◯ *Report*	

Visual Inspection	OK	NOT OK	List All Corrective Work Done **AND** report all Work Notifications written
	◯	◯ *Report*	

Procedure Specific Tasks	OK	NOT OK	List All Corrective Work Done **AND** report all Work Notifications written.
As-Found	◯	◯	
As Left	◯	◯ *Report*	

This PM performed by _____ on _____ .

Manufacturer		Model and Type		Specific Equip. Data
Equipment Type		Serial No.		
Application		Location		
Drawing Number				
SAP Identifier				
Spare		*Spare*		

Validate Specific Equipment Data	OK	NOT OK	List Missing or Incorrect Data
As Found	O	O Report	

Visual Inspection	OK	NOT OK	List All Corrective Work Done **AND** report all Work Notifications written
	O	O Report	

Procedure Specific Tasks	OK	NOT OK	List All Corrective Work Done **AND** report all Work Notifications written
As-Found	O	O	
As Left	O	O Report	

This PM performed by	_____	on	_____

Manufacturer		Model and Type	Specific Equip. Data
Equipment Type	Electrical Panel		
Purpose(s)	1. Maintain component reliability, improve safety and increase production by eliminating unplanned downtime through properly planned and scheduled maintenance.		

Before proceeding with this work, the equipment needs to be safe. This means that all potential energy sources, such as pressure or power, need to be positively identified and locked out/tagged out according to your platform specific LO/TO procedures for this of equipment.

Take time to plan and think about your work.

Inform Operations (Platform Lead, PRC, or Field Foreman) before proceeding with the following maintenance activity.

For shutdown PM's, confirm that all potential energy sources have been identified, and LO/TO.

For all PM's, be aware of the other activities going on around you, and take necessary precautions to eliminate maintenance-induced failures.

Consumables	
Spare Parts	Part Number

Step	Action	Notes, Diagrams and Key Notes
A *Preparation* *Visual*	1. Ensure system is safe to work on **OR** Ensure system is properly LO/TO at panel board. 2. Perform visual inspection (look for leaks, irregular sounds/noise, abnormalities and loose components). 3. Review drawings and identify relevant components. 4. Inform Production Operations of the tests to be performed.	

B-1		Note :
Task	1) Remove any unnecessary clutter both inside and outside of panel location.	
	2) Inspect panel wall mounting bolts for tightness.	
	3) Clean both outside and inside panel,	
	4) wipe with lint free rag or vacuum if needed.	
	5) Actuate on/off type switches to verify proper movement and contact connection.	
	6) Inspect all (6 total) indication lights to verify the bulbs are not burnt out and the lens covers are not broken and in place.	
	7) Verify all cabling and conduit connections are secure and properly terminated.	
	8) Inspect all power and control wiring terminations for tightness.	
	9) Inspect wiring for any signs of deterioration or excessive heating in the insulation.	

B-2	If ...	Then ...	Note :
		1.	

	If ...	Then ...
B-3		
B-4		
B-5	1.	
C-1 *Return to Service*	1. When completed inform Production Operations that equipment is in service.	
C-2 *Clean Up*	1. Clean up and remove all debris, parts and/or tools. 2. Dispose of discarded parts and/or contaminated parts as per HSE procedures.	
D *Data*	Complete information where indicated on the Notification and Feedback form.	*Accurate reporting creates accurate equipment histories*

Notification Name	Start Air Compressor Oil Level
Location	CBE-900 Gas Compressor
Drawing Number	
Equipment Application	• CBE-900 Starting Air Compressor • CBE-901 Starting Air Compressor • CBE-902 Starting Air Compressor
Frequency	Daily
Priority	1-Operations Monitoring
Craft	Operator
Procedure Ref.	CPR-Q-012-P
Time	0.25 Hour

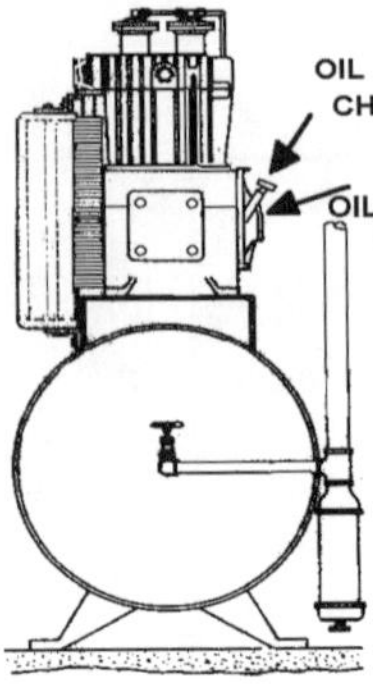

OIL LEVEL CHECK

OIL PRESSURE CHECK

Validate Equipment Data

As Found: OK ☐

                NOT OK ☐

• CBE-900 Gas Compressor

If not correct record actual data:

Start Air System Visual Inspection

As Found: OK ☐

                NOT OK ☐

List Any Corrective Action Performed (use back of sheet if necessary)

W.R.# _____

Lube Oil Level

CBE-900 added:_____qt

                OK ☐

CBE-901 added:_____qt

                OK ☐

CBE-902 added:_____qt

                OK ☐

List Any Corrective Action Performed

W.R.# _____

This PM performed by _____

Today's date _____

Figure C-2 Compressor: Daily Oil Check

Notification Name	Gas Engine/Compressor Daily Check
Location	Gas Compressor
Drawing Number	
Equipment Application	• CBE-900 Compressor/Engine
Frequency	Daily
Priority	1-Operations Monitoring
Craft	OPS
Procedure Ref.	CBA-C-010-P (no data found)
Time	1 Hour

Validate Equipment Data
As Found: OK ☐
              NOT OK ☐

• CBE-900 Compressor/ Engine
If not correct record actual data:

Gas Engine System Visual Inspection
As Found: OK ☐
              NOT OK ☐

List Any Corrective Action Performed
(use back of sheet if necessary)

W.R.# _____

Perform power cylinder balancing
                                ☐

List Any Corrective Action Performed

W.R.# _____

Continued on next page

Figure C-3 Compressor: Daily

CYL	^{0}F	CYL	^{0}F
1		7	
2		8	
3		9	
4		10	
5		11	
6		12	

Monitor engine cylinder temperatures: ☐

Write-in actual measurements in table below:

List Any Corrective Action Performed

W.R.# _____

Monitor lube oil level:

Level OK ☐

Added: _____ quarts ☐

Type of oil:

Shell _____ (SAE____)

W.R.# _____

This PM performed by _____

Today's date _____

Notification Name	Gas Compressor/Engine Monthly Checks
Location	CBE-900 Gas Compressor
Drawing Number	
Equipment Application	• CBE-900 Compressor/ Engine
Frequency	Monthly
Priority	1-Operations Monitoring
Craft	OPS
Procedure Ref.	Compressor Manual – In Maintenance Library (see Thermographic instruction manual) (SOI: Oil Sampling Program)
Time	1 Hour

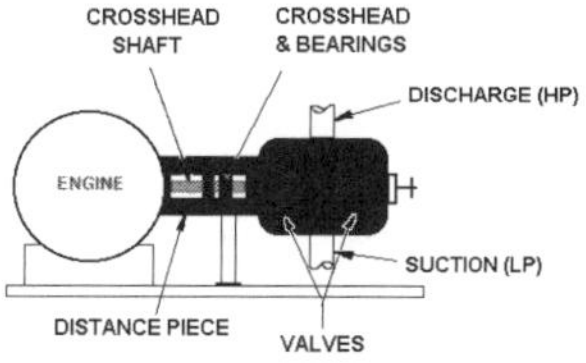

Validate Equipment Data **As Found:** OK ☐ NOT OK ☐	• CBE-900 Compressor/ Engine **If not correct record actual data:**
Engine/ Compressor System Visual Inspection **As Found:** OK ☐ NOT OK ☐	**List Any Corrective Action Performed (use back of sheet if necessary)** **W.R.#** _____ NOTE: **If shaft packing or valve leakage is identified, initiate Work Request for required maintenance.**

Continued on next pages

Figure C-4 Compressor: Monthly

Perform thermography on compressor:	List Any Corrective Action Performed
1st Stage ☐ 2nd Stage ☐ 3rd Stage ☐	 W.R.# _____
Perform lube oil sampling: Engine ☐ Compressor ☐	List Any Corrective Action Performed NOTE: After taking sample send it to outside vendor for analysis. W.R.# _____ P.O.# _____

This PM performed by   _____

Today's date   _____

Notification Name	Gas Compressor/ Engine Quarterly Checks
Location	CBE-900 Gas Compressor
Drawing Number	333C-91-2S
Equipment Application	• CBE-900 Compressor/ Engine
Frequency	Quarterly
Priority	2-Scheduled Maintenance
Craft	Mechanical
Procedure Ref.	(see BETA instructions)
Time	2 Hours

Validate Equipment Data

As Found: OK ☐

            NOT OK ☐

• CBE-900 Compressor/ Engine

If not correct record actual data:

Engine/ Compressor System Visual Inspection

As Found: OK ☐

            NOT OK ☐

List Any Corrective Action Performed (use back of sheet if necessary)

W.R.# _____

NOTE: Replace turbocharger if there are signs of degradation.

Perform BETA Analysis

            Turbocharger ☐

            1st Stage Compressor ☐

            2nd Stage Compressor ☐

            3rd Stage Compressor ☐

Check engine ignition timing

            OK   ☐

            adjusted   ☐

List Any Corrective Action Performed

NOTE: Adjust or service as required. List any corrective action performed.

W.R.# _____

Figure C-5 Compressor: Quarterly

Notification Name	Change Compressor Crankcase Oil
Location	CBE-900 Gas Compressor
Drawing Number	
Equipment Application	• CBE-900 Air Compressor • CBE-901 Air Compressor • CBE-902 Air Compressor
Frequency	Quarterly
Priority	2-Lubrication
Craft	Mechanical
Procedure Ref.	Compressor Manual
Time	1 Hour (each compressor)

Validate Equipment Data **As Found:** OK ☐ NOT OK ☐	• CBE-900 Gas Compressor **If not correct record actual data:**
Start Air System Visual Inspection **As Found:** OK ☐ NOT OK ☐	**List Any Corrective Action Performed** **(use back of sheet if necessary)** **W.R.#** _____
Changed crankcase oil: **CBE-900** ☐ **CBE-901** ☐ **CBE-902** ☐	**List Any Corrective Action Performed** **W.R.#** _____

✎ **This PM performed by** _____
   Today's date _____

✎
✎
✎

Figure C-6 Compressor: Quarterly Oil Change

Notification Name	Gas Compressor/ Engine Semi-Annual Checks
Location	CBE-900 Gas Compressor
Drawing Number	
Equipment Application	• CBE-900 Compressor/ Engine
Frequency	Semi-Annual
Priority	2-Scheduled Maintenance
Craft	Electrical
Procedure Ref.	CBA-C-013-P
Time	4 Hours

Validate Equipment Data As Found: OK ☐ NOT OK ☐	• CBE-900 Compressor/ Engine If not correct record actual data:
Engine/ Compressor System Visual Inspection As Found: OK ☐ NOT OK ☐	List Any Corrective Action Performed (use back of sheet if necessary) W.R.# _____
• Replace all high-tension (unsheilded) leads. ☐ • Replace all spark plugs. ☐ • Change air intake filter elements. ☐	List Any Corrective Action Performed W.R.# _____

✍ This PM performed by   _____

    Today's date   _____

✍

✍

Figure C-7 Compressor: Semi-Annual

Notification Name	Air Compressor Monthly Inspection
Location	CBE-900 Compressor
Drawing Number	
Equipment Application	• CBE-880 Start Air Compressor • CBE-901 Start Air Compressor • CBE-902 Start Air Compressor
Frequency	Monthly
Priority	2-Scheduled Maintenance
Craft	Operator
Procedure Ref.	
Time	1 Hour

Validate Equipment Data As Found:　　OK ☐ 　　　　NOT OK ☐	• CBE-900 Gas Compressor If not correct record actual data:
Start Air System Visual Inspection As Found:　　OK ☐ 　　　　NOT OK ☐	**List Any Corrective Action Performed** **(use back of sheet if necessary)** W.R.# _____

Continued on next pages

Figure C-8 Compressor: Monthly Operator

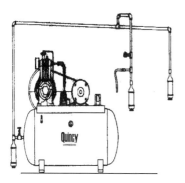

✎ Intake air filter element:		List Any Corrective Action Performed
1. Cleaned	☐	
or 2. Replaced	☐	
After Cooler:		
Cleaned surfaces	☐	
Intercooler surfaces	☐	
Inspected Oil:		
OK	☐	Note:
Not OK	☐	Change oil on PM-001-001
Check drive belt:		Note:
OK	☐	If "Not OK" write a work request to change belt.
Not OK	☐	
		W.R.# _____

✎ This PM performed by _____

Today's date _____

Gas Compressor Weekly

1. SAFETY:
   1.1 Contractor personnel must sign in at Safety office located in the gas plant control building.
   1.2 Obtain proper documentation and signatures from operations prior to beginning any maintenance.
   e.g.: Vehicle entry, Safe work, Hot work, Confined space, etc.

2. EQUIPMENT:
   2.1 Clean lint free rags
   2.2 Approved non-flammable solvent
   2.3 Straight edge

3. PARTS:
   3.1 Oil, Mobil 485

4. PROCEDURES:
   4.1 Use good housekeeping practices while performing maintenance.
   4.2 Pressure Vessel inspection:
       4.2.1 Check suction/discharge of vessel and associated piping for evidence of leaks and vibration. If leaks are detected, generate a work order for corrective action. If vibration is detected at brace or support areas immediately contact shift maintenance supervisor.
       4.2.2 Visually inspect the externals of the vessel for evidence of leaks or corrosion at manhole/flanges. If leaks are detected, generate a work order for corrective action.
   4.3 Engine inspection:
       4.3.1 Using the Manometer located on the west wall under the air pipeline, check the pressure drop across the engine air filters. Clean the filters if the manometer reading is higher than 2" H2o.
       4.3.2 Check the pressure readings across the following gauges and generate a work order if the pressure drop exceeds the listed tolerances:
           4.3.2.1 Turbocharger oil filter, located at inlet and outlet sides of the turbo oil filter supply line.   15 psi max.   actual_____
           4.3.2.2 Engine oil filter and strainer, located on the inlet and outlet sides of the oil supply at the filter box.   15 psi max.   actual_____
       4.3.3 Visually inspect the engine for inappropriate equipment conditions. Check for air fuel, or lube oil leaks. Generate a work order for any corrective action.
   4.4 Compressor inspection:
       4.4.1 Check compressor cylinder lubricating block for proper operation.
           4.4.1.1 Procedure to be outlined
       4.4.2 Check lubricator lines for leaks and appropriate condition.
           4.4.2.1 Procedure to be outlined
       4.4.3 Using the dipstick determine the crankcase oil level. Oil level must be within the defined area of the dipstick. If the oil level is low fill to appropriate level with Mobil 485 oil.
       4.4.4 Verify correct operation of lubricator pump and outputs.
           4.4.4.1 Using primer buttons verify oil output from lubricators.
           4.4.4.2 Clean the lubricator reservoir sight glass with approved solvent and clean rag.
           4.4.4.3 Oil level in the lubricator reservoir must be within the sight glass. If low, fill to appropriate level with Mobil 485.
   4.5 Pump inspection:
       4.5.1 Visually inspect the engine pre-lube pump for leaking packing glands. Generate a work order for any corrective action.
       4.5.2 Using a straight edge verify pump and motor alignment. Generate a work order for any corrective action.

continued on next page

Figure C-9 Gas Compressor: Detailed Service

4.5.2.1 To determine alignment, place a straight edge diametrically across the outside face of the driven shaft and bring the straight edge to line up diametrically with the outside face of the driving shaft. If the components are correctly aligned to each other the straight edge will lie squarely on both shaft faces. If the straight edge does not lie square, the driving component must be move until this is accomplished. Identical spacer blocks may be necessary in-between the shafts and the straight edge to allow clearance for the coupling.

4.6 Electrical inspection:

4.6.1 Visually inspect the external of motors for inappropriate conditions. Check for loose base bolts, vent ducts clean an free of obstruction, unusual heat buildup. Generate a work order for any corrective action.

4.7 Fin Fan Inspection:

4.7.1 Visually inspect externals for leaks, corrosion, and discoloration. If conditions are found, generate work order for corrective action.

4.7.2 Visually inspect suction/discharge of fin fan and associated piping for leaks and vibration. If leaks are detected, generate a work order for any corrective action. If vibration is detected at brace or support area immediately contact the shift maintenance supervisor.

4.7.3 Check the temperature gauges at the inlet and outlet ports to verify correct operating temperature. If temperature is found out of perimeters generate a work order for cleaning.

Inlet - 195degrees to 225 degrees	actual_____
Outlet - 180 degrees to 190 degrees	actual_____

4.7.4 Check the pressure gauges at the inlet and outlet ports to verify correct operating pressure. If pressure is found out of perimeters generate a work order for cleaning.

Inlet - 40psi to 45psi	actual_____
Outlet - 40psi to 45psi	actual_____

4.8 Post inspection:

4.8.1 Upon completion of maintenance inform operations of completion. Ensure all necessary work orders are generated. Get shift maintenance supervisor to sign work completion and close work order.

Gear Box Inspection

1.0 PURPOSE: This procedure provides detailed instructions for inspecting the gears to determine the rate of wear by measuring backlash and total indicated runout of gear teeth crowns.

2.0 SCOPE: This procedure supplements the existing daily PM inspection requirements.

3.0 EQUIPMENT:
    3.1 Mechanic tools
    3.2 Portable lamp / Flashlight
    3.3 Inspection mirror
    3.4 Thickness gauges / Plastigauge
    3.5 Dial indicator with magnetic post stand
    3.6 Approved cleaning solvent
    3.7 Lint free cleaning rags

4.0 SAFTEY:
    4.1 Ensure coordination with operations
    4.2 Observe site, and area safety precautions at all times
    4.3 Ensure equipment is isolated from all power sources and all LO/TO/TO procedures are completed
    4.4 Ensure that the work area is cleared of all debris and foreign material before opening inspection ports
    4.5 Remove objects from pockets that could fall into gear box while inspection ports are open

5.0 PROCEDURE:
    5.1 Clean surface area around inspection doors with approved cleaning solvent and lint free rags.
    5.2 Loosen and remove inspection door fasteners.
    5.3 Remove inspection door and place out of the way in a clean location.
    5.4 Remove gasket/O-ring. Ensure no foreign matter falls into the gear box during door removal.

CAUTION: *Foreign matter introduced into the gear box can result in severe damage to the gear teeth or foul the lubrication system.*

    5.5 Inspect gasket for any deformities or abnormal crush pattern. NOTE: *Do not discard the gasket at this time*
    5.6 Ensure no abnormal surface conditions exist on inspection door and gear box casing at gasket contact points.
    5.7 Clean gasket surface area. Remove any adhesions that could cause oil leaks during operation.
    5.8 Visually inspect condition of inspection door bolts and nuts. Replace any found damaged.
    5.9 Inform operations that the equipment will be rolled by hand to perform gear back lash measurements.
    5.10 Use a pinch bar to roll the equipment until the drive face of the gear teeth are firmly meshed together. CAUTION: *Keep fingers and materials clear of gear meshing areas.*
    5.11 Clean inspection areas of gear teeth with lint free rag soaked in approved cleaning solvent. Wipe inspection area dry after cleaning is complete.

    5.12 Use plastigage or certified thickness gauge to measure the backlash clearance. Take the measurement at a radius approximately equal to the pitch radius of the pinion. This gives a direct reading of backlash. NOTE: *The specified backlash clearance is 0.018" - 0.027"* Record the clearance measurement.

continued on next page

Figure C-10 Gear Box: Detailed Inspection

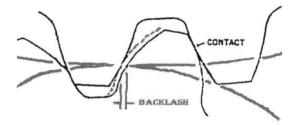

5.13   Inform your supervisor if the actual measurement is out of the accepted tolerance. Wait for an engineering decision before continuing with the inspection.

5.14   Proceed with the inspection if the clearance is within the acceptable tolerance.

5.15   Initiate check for gear eccentricity and inspection of gear teeth. Inform operations this process will require the asset to be rolled slowly for a few revolutions to inspect teeth condition.

5.16   Mount dial indicator magnetic stand on the gear box casing. Set indicator stem on crown of gear tooth. Zero the indicator after the needle has made one revolution to ensure the indicator does not bottom out.

5.17   Slowly roll the unit and take dial indicator measurements at approximately 90 degree intervals and record readings.

5.18   Inspect teeth faces for wear patterns such as:

5.18.1   WEAR - Any loss of tooth material from the contacting surfaces of gear teeth.

5.18.2   ABRASION - Tooth surface injury caused by fine particles passing through the gear mesh.

5.18.3   SCRATCHING - Severe form of abrasive wear showing short, scratch like marks on contact surface areas in the direction of sliding.

5.18.4   ROLLING and PEENING - Sharp fin like ridges on the edges or tops of the teeth. This is caused by impact loading and severe sliding due to excessive loading. Do not confuse these with machining burrs.

5.18.5   OVERLOAD - Removal of metal in thin layers or flakes which leaves etching like markings on the contact surfaces of the teeth.

5.18.6   RIPPLING - A wave-like formation on the tooth surface at right angles to the direction of sliding. It is characterized by fish scale appearance and occurs mainly on case hardened hypoid pinions.

5.18.7   SCORING - A rapid removal of metal from the tooth surfaces. The scored surface is identified by a torn or dragged and furrowed appearance in the direction of sliding. It is caused by the tearing out of small contacting particles that have welded together as a result of metal-to-metal contact.

5.19   Inform your supervisor if the gear teeth show any of the fore mentioned signs of wear. Wait for an engineering decision before continuing with the inspection.

5.20   Replace inspection door with a new gasket.

5.21   Clean up entire work area, and account for all tools used. Return tools to tool crib.

5.22   Inform operations repair is complete, close out work order.

Notification Name	Lube Oil Level Controller Oil Level
Location	CBE-900 Gas Compressor
Drawing Number	
Equipment Application	Engine Lube Oil Level Controller
Frequency	Daily
Priority	1-Operations Monitoring
Craft	Operator
Procedure Ref.	None
Time	0.5 Hour

Validate Equipment Data
As Found: OK ☐
NOT OK ☐

- Engine Lube Oil Level Controller
- **If not correct record actual data:**

Lube Oil System Visual Inspection
As Found: OK ☐
NOT OK ☐

List Any Corrective Action Performed (use back of sheet if necessary)

W.R.# _____

Engine Lube Oil Level
OK ☐
Not OK ☐

List Any Corrective Action Performed

If "not ok" write a W.R. to check "level controller function and calibration."

W.R.# _____

✍ **This PM performed by** _____
Today's date _____

✍
✍
✍

Figure C-11 Lube Oil: Daily

Notification Name	Lube Oil Level Controller Sediment Bowl
Location	CBE-900 Gas Compressor
Drawing Number	
Equipment Application	• Engine Lube Oil Level Controller
Frequency	Semi-Annual
Priority	2-Scheduled Maintenance
Craft	Mechanical
Procedure Ref.	
Time	0.5 Hour

Validate Equipment Data **As Found:** **OK** ☐ **NOT OK** ☐	**• Lube Oil Level Controller** **If not correct record actual data:**
Lube Oil System Visual Inspection **As Found:** **OK** ☐ **NOT OK** ☐	**List Any Corrective Action Performed** **(use back of sheet if necessary)** **W.R.#** _____
Cleaned sediment bowl & screen. **OK** ☐	**List Any Corrective Action Performed** **W.R.#** _____

✍   This PM performed by   _____

     Today's date   _____

✍

✍

✍

Figure C-12 Lube Oil: SA

\<Motor Control Center\>

Equipment Data		
Production Unit	Service	
Field	System	Electrical
Platform	Component	Motor Control Center

PM Data		
Frequency	Annual	Craft	Electrical
Priority		Shutdown Req ?	**Yes**
Use Procedure		Total Duration	4 hour
Procedure file		MMS ?	**No**

Before proceeding with this work, the equipment needs to be safe. This means that all potential energy sources, such as pressure or power, need to be positively identified and locked out/tagged out according to your platform specific LO/TO procedures for this of equipment.

Equipment List		
SAP Identifier		
SAP Identifier		

Figure C-13 Motor Control Center Electrical: Annual

Manufacturer	Allen Bradley	Model and Type		Specific Equip. Data
Equipment Type	Motor Control Center	Serial No.		
Application		Location	Motor Control Center	
Drawing Number				
SAP Identifier				
Spare		Spare		

Validate Specific Equipment Data	OK	NOT OK	List Missing or Incorrect Data
As Found	O	O *Report*	

Visual Inspection	OK	NOT OK	List All Corrective Work Done **AND** report all Work Notifications written
	O	O *Report*	

Procedure Specific Tasks	OK	NOT OK	List All Corrective Work Done **AND** report all Work Notifications written.
As-Found	O	O	
As Left	O	O *Report*	

This PM performed by _____ on _____.

	Specific Equip. Data
Manufacturer	
Equipment Type	Model and Type
Application	Serial No.
Drawing Number	Location
SAP Identifier	
Spare	Spare

Validate Specific Equipment Data	OK	NOT OK	List Missing or Incorrect Data
As Found	O	O Report	

Visual Inspection	OK	NOT OK	List All Corrective Work Done AND report all Work Notifications written
	O	O Report	

Procedure Specific Tasks	OK	NOT OK	List All Corrective Work Done AND report all Work Notifications written.
As-Found	O	O	
As Left	O	O Report	

This PM performed by _____ on _____ .

Motor Control Center

		Specific Equip. Data
Manufacturer	Allen Bradley	
	Model and Type	
Equipment Type	Electrical Switches	
Purpose(s)	1. Maintain component reliability, improve safety and increase production by eliminating unplanned downtime through properly planned and scheduled maintenance.	

Before proceeding with this work, the equipment needs to be safe. This means that all potential energy sources, such as pressure or power, need to be positively identified and locked out/tagged out according to your platform specific LO/TO procedures for this of equipment.

Take time to plan and think about your work.

Inform Operations (Lead or Foreman) before proceeding with the following maintenance activity.

For shutdown PM's, confirm that all potential energy sources have been identified, and LO/TO.

For all PM's, be aware of the other activities going on around you, and take necessary precautions to eliminate maintenance-induced failures.

		Requirements
Tools	A Hand tools. Volt/amp meter	
Consumables		
Spare Parts		*Part Number*

Step	*Action*	*Notes, Diagrams and Key Notes*
A *Preparation Visual*	1. Ensure system is safe to work on **OR** Ensure system is properly LO/TO at panel board. 2. Perform visual inspection (look for leaks, irregular sounds/noise, abnormalities and loose components). 3. Review drawings and identify relevant components. 4. Inform Production / Operations of the tests to be performed.	

B-1	
Task	*Notes :*

1. Remove any excessive clutter around motor control center.
2. Inspect all incoming cables for proper cable connections at MCC unit or transition piece. Verify that there is no possible way for water to leak inside and down cables. Seal as needed with a water proof sealant
3. Verify all sections are identified with a nametag.
4. Inspect foundation for any obvious problems. Verify all anchor bolts are in place and tight.
5. Inspect main ground wire for any excessive corrosion around lug and connection point to MCC ground buss bar. Verify ground wire is securely terminated.
6. Inspect all buss joints or ties for tightness, torque to manufacture specifications.
7. Check incoming power connections to buss bars. Torque bolts to manufacturer's specifications
8. Remove plug-in units, check stab assemblies for proper alignment and contact.
9. For each starter section:
    a. Remove any accumulated dust, spare parts, etc. from each section.
    b. Inspect all power and control wiring terminations, for tightness and any signs of deterioration of insulation.
    c. Inspect fuses and overload heater for proper sizing. Field verify, compare to equipment nameplate rating.

rating.

11. Inspect the 32volt heater circuit. Verify control fuses are 600volt 15amps. Verify heater is working by performing voltage

12. Measure insulation resistance of each buss section, phase to phase and phase to ground for one minute----record values.

13. Measure insulation resistance of each starter section phase to phase and phase to ground with the starter contacts closed and protective device opens.

14. On pump control panel:

   a. Inspect all indication lamps. Verify all are functional with no burnt out bulbs or missing lamp covers.

   b. Inspect all other controls such as selector switches and pushbuttons to verify proper operation.

   c. Remove any unnecessary clutter from within panel.

   d. Inspect all power and control wiring terminations for tightness.

   e. Verify all timers for proper setting. Refer to wiring diagrams.

   f. Inspect inside of cabinet for any signs of moisture or corrosion.

   g. Verify proper control fuses rating. Refer to wiring diagrams

15. Verify a current set of diagrams is present for MCC unit.

16. Perform a Infrared scan on all MCC components on a bi-annual frequency.

B-2

If …	Then …
	1. . .

Note :

B-3		
B-4	If ...	Then ...
B-5	1.	
C-1 *Return to Service*	1. When completed inform Production Operations that equipment is in service.	
C-2 *Clean Up*	1. Clean up and remove all debris, parts and/or tools. 2. Dispose of discarded parts and/or contaminated parts as per platform HSE procedures.	
D *Data*	Complete information where indicated on the Notification and Feedback form.	*Accurate reporting creates accurate equipment histories*

Installing Gland Packing in Centrifugal Pumps

1.0 PURPOSE: This procedure provides detailed instruction on repacking centrifugal pump glands. This methodology is applicable to other gland sealed units such as valves and reciprocating machinery.

2.0 SCOPE   This procedure supplements existing site procedures.

3.0 EQUIPMENT:
    3.1   Mechanic's Tools
    3.2   Approved packing for specific asset, Store number #####
    3.3   Mandrel sized to shaft diameter, # #/#inch
    3.4   Packing ring extractor tool
    3.5   Packing board and sharp knife
    3.6   Approved cleaning solvent
    3.7   Lint free cleaning rags

4.0 SAFEY
    4.1   Coordinate repair outage with operations control.
    4.2   Observe site, an area safety precautions at all time.
    4.3   Ensure asset has been isolated from all power sources and L.O. T.O. T.O. procedures have been completed.
    4.4   Ensure pump is isolated, and depressurized with suction and discharge valves chained and locked closed.

5.0 PROCEDURE
    5.1   Loosen and remove nuts from the gland bolts.
    5.2   Examine threads on bolts and nuts for stretching or damage - replace if defective.
    5.3   Remove the gland follower from the stuffing box and slide it along the shaft for access to the packing area.
    5.4   Use packing extraction tool to carefully remove packing from the gland.
    5.5   Retain each packing ring in the order that it is removed from the gland box for examination of wear characteristics. Look for rub marks and any other unusual markings that would identify potential repair problems.
    5.6   Carefully remove the lantern ring. (This is a grooved, bobbin-like spool piece that is situated exactly on the center-line of the seal water inlet connection to the gland.) NOTE: *It is important to locate the lantern ring under the seal water inlet connection to ensure the seal water is properly distributed within the gland to perform its cooling and lubricating functions.*
    5.7   Examine the lantern ring for scoring and possible signs of crushing. Check that the ring's outside diameter is equal to a sliding fit inside the gland box internal diameter. Check the ring's inside diameter is a free fit along the pump's shaft sleeve. If any tolerance is found unacceptable replace the lantern ring with a new ring.
    5.8   Continue to remove the rest of the packing rings as previously described. Retain each ring in the sequence that it was removed. Thoroughly examine the rings for potential problem signs and causes.
    5.9   Turn on the gland seal cooling water slightly to ensure there is no blockage in the line. Shut the valve when good coolant flow is established.
    5.10 Repeat procedural steps 5.1 through 5.9 for the second gland box.
    5.11 Clean out the gland stuffing boxes carefully with a solvent soaked cleaning rag, ensuring no debris are left behind.
    5.12 Examine the shaft sleeve in both gland areas for wear that may be caused by poorly lubricated or over tightened packing. NOTE: *If the shaft sleeve is grooved or badly scratched in any way, the split casing of the pump will have to be split to remove the impeller for the sleeve to be replaced.*

Figure C-14 Pump Packing

5.13 Check the Total Indicated Runout of pump shaft by placing a magnetic base mounted dial indicator on the pump casing and dial stem on the pump shaft. Zero the dial and rotate the pump shaft one full turn. Record the reading. NOTE: *If the TIR is greater than +/- 0.002" the pump shaft must be straightened*

5.14 Determine the correct size of packing before installing. Determine the correct packing size as follows:
Outside diameter of the gland opening minus the shaft diameter divided by 2
NOTE: *Use only the correct size packing, which is available in increments of 1/32 of an inch. Controlled leakage can easily be achieved with the correct size packing.*

5.15 Cut the packing rings to size on a wooden mandrel that is the same diameter as the pump shaft. Cut the packing rings diagonally at approximately 30^0. NOTE: *Leave 1/16 inch gap between the butts of the packing ring. This permits the packing rings to move under compression or temperature with out binding on the surface of the shaft.*

5.16 Ensure the gland area is perfectly clean and is not scratched in any way before installing the packing rings.

5.17 Lubricate each ring lightly before installing into the gland stuffing box. NOTE: *When putting the packing rings around the shaft use a "S" twist. DO NOT BEND OPEN.*

5.18 Use a split bushing to install each ring. Ensure each ring bottoms out inside the stuffing box. DO NOT USE A SCREWDRIVER.

5.19 Stager the butt joints in the following pattern:

5.19.1   First ring butts at 12 o'clock
5.19.2   Second ring butts at 6 o'clock
5.19.3   Third ring butts at 3 o'clock
5.19.4   Fourth ring butts at 9 o'clock
5.19.5   Repeat this sequence until the gland has been filled

NOTE: *When the last ring has been installed, there should be enough room to insert the gland follower - 1/8 to 3/16 inch.*

5.20 Install the lantern ring in its correct position with in the glad. Do not force the lantern ring in to position.

5.21 Tighten the gland bolts with a wrench to seat and form the packing to the stuffing box and shaft.

5.22 Loosen the gland nuts one complete turn and rotate the shaft by hand to get running clearance.

5.23 Re-tighten the nuts hand tight only. Rotate the shaft by hand again to ensure the packing is not too tight.

5.24 Start the pump and allow the stuffing box to leak, then tighten the gland bolts one flat at a time until the leakage is controlled. Exercise caution not to over tighten as this will cause the pump to run hot.

5.25 Clean up entire work area, and account for all tools used. Return tools to tool crib.

5.26 Inform operations repair is complete, close out work order.

5.27 After the pump is in operation periodically inspect the gland to determine its performance. If the seal is found to leak tighten the gland bolts one flat at a time. Give the packing time to adjust before tightening it more. If the gland is tightened too much at one time it can compress the packing excessively causing unnecessary friction and subsequent packing burn out.

Switchboard #1

Equipment Data

Production Unit	Service	Turbine Generator
Field	System	
Platform	Component	Control Panel

PM Data

Frequency	Quarterly	Craft	Electrical
Priority		Shutdown Req ?	No
Use Procedure		Total Duration	1 hour
Procedure file		MMS ?	No

Before proceeding with this work, the equipment needs to be safe. This means that all potential energy sources, such as pressure or power, need to be positively identified and locked out/tagged out according to your platform specific LO/TO procedures for this of equipment.

Equipment List

SAP Identifier		
SAP Identifier		

Figure C-15 Switchboard: Electrical

Specific Equip. Data			
Manufacturer	Model and Type		
Equipment Type	Electrical Panel	Serial No.	
Application		Location	Switchgear Room
Drawing Number			
SAP Identifier			
Spare		Spare	

Validate Specific Equipment Data	OK	NOT OK	List Missing or Incorrect Data
As Found	○	○ Report	

Visual Inspection	OK	NOT OK	List All Corrective Work Done **AND** report all Work Notifications written
	○	○ Report	

Procedure Specific Tasks	OK	NOT OK	List All Corrective Work Done **AND** report all Work Notifications written.
As-Found	○	○	
As Left	○	○ Report	

This PM performed by _____ **on** _____ .

		Specific Equip. Data
Manufacturer	Model and Type	
Equipment Type	Serial No.	
Application	Location	
Drawing Number		
SAP Identifier		
Spare	Spare	

Validate Specific Equipment Data	OK	NOT OK	List Missing or Incorrect Data
As Found	○	○ *Report*	

Visual Inspection	OK	NOT OK	List All Corrective Work Done **AND** report all Work Notifications written
	○	○ *Report*	

Procedure Specific Tasks	OK	NOT OK	List All Corrective Work Done **AND** report all Work Notifications written.
As-Found	○	○	
As Left	○	○ *Report*	

This PM performed by _____ **on** _____ .

Manufacturer		Model and Type	Specific Equip. Data
Equipment Type	Electrical Panel		
Purpose(s)	1. Maintain component reliability, improve safety and increase production by eliminating unplanned downtime through properly planned and scheduled maintenance.		

Before proceeding with this work, the equipment needs to be safe. This means that all potential energy sources, such as pressure or power, need to be positively identified and locked out/tagged out according to your platform specific LO/TO procedures for this of equipment.

Take time to plan and think about your work.

Inform Operations (Lead or Foreman) before proceeding with the following maintenance activity.

For shutdown PM's, confirm that all potential energy sources have been identified, and LO/TO.

For all PM's, be aware of the other activities going on around you, and take necessary precautions to eliminate maintenance-induced failures.

	Requirements	
Tools	➤ Hand tools. Volt/Amp Meter	
Consumables		
Spare Parts		Part Number

Step	Action	Notes, Diagrams and Key Notes
A *Preparation* *Visual*	1. Ensure system is safe to work on **OR** Ensure system is properly LO/TO at panel board. 2. Perform visual inspection (look for leaks, irregular sounds/noise, abnormalities and loose components). 3. Review drawings and identify relevant components. 4. Inform Production Operations of the tests to be performed.	
B-1 *Task*	1. Remove any unnecessary clutter around switchboard control panel. 2. Inspect foundation and structural support for any signs of obvious concerns. Verify support bolts are not missing and tight. 3. Verify the external ground wire is properly terminated to the switchboard enclosure frame. 4. Verify the front panel mounted gauges are operational. They include the AC Volts, Hertz, AC Kilowatts, and Remove any unnecessary clutter around generator and AC Amperes. Verify accuracy with a Fluke meter.	<u>*Note :*</u>

5. Inspect all indicator lights for proper operation with no burnt bulbs or missing lamp covers.

6. When possible actuate all other switches to verify smooth operation and proper contact closure.

7. Inspect all power and control wiring terminations for tightness.

8. Verify all agastat's are properly set. Refer to schematic diagrams.

9. Verify all control fuses are properly sized. Refer to schematic diagrams.

10. Using a voltmeter verify proper voltage on control transformer. 480v primary and 120v secondary.

11. Using a megohm meter "meg" power leads from switchboard to the generator. If results are above 1.5 megohms, follow up action is needed.

12. Verify a current copy of the wiring schematic exist in the panel.

B-2	If ...	Then ...
		1. . .

Note :

B-3		
B-4	If ...	Then ...
B-5	1.	
C-1 Return to Service	1. When completed inform Production Operations that equipment is in service.	
C-2 Clean Up	1. Clean up and remove all debris, parts and/or tools. 2. Dispose of discarded parts and/or contaminated parts as per platform HSE procedures.	
D Data	Complete information where indicated on the Notification and Feedback form.	*Accurate reporting creates accurate equipment histories*

Miscellaneous Starters/Contactors

Equipment Data	
Production Unit	
Field	
Platform	

	Service	
	System	Electrical
	Component	Starter or contactor

PM Data	
Frequency	Bi-annual
Priority	
Use Procedure	
Procedure file	

	Craft	Electrical
	Shutdown Req ?	**Yes**
	Total Duration	1/2 hour each
	MMS ?	**No**

Before proceeding with this work, the equipment needs to be safe. This means that all potential energy sources, such as pressure or power, need to be positively identified and locked out/tagged out according to your platform specific LO/TO procedures for this of equipment.

Equipment List			
SAP Identifier			
SAP Identifier			

Figure C-16 Switches and Starters

Manufacturer	Various			Specific Equip. Data
Equipment Type	Electrical switches		Model and Type	
Application			Serial No.	
Drawing Number			Location	Compressor Room
SAP Identifier				
Spare		*Spare*		

Validate Specific Equipment Data	OK	NOT OK	List Missing or Incorrect Data
As Found		*Report*	

Visual Inspection	OK	NOT OK	List All Corrective Work Done **AND** report all Work Notifications written
		Report	

Procedure Specific Tasks	OK	NOT OK	List All Corrective Work Done **AND** report all Work Notifications written.
As-Found			
As Left		*Report*	

 **This PM performed by** _____ on _____ .

			Specific Equip. Data
Manufacturer	Model and Type		
Equipment Type	Serial No.		
Application	Location		
Drawing Number			
SAP Identifier			
Spare	Spare		

Validate Specific Equipment Data	OK	NOT OK	List Missing or Incorrect Data
As Found	O	O Report	

Visual Inspection	OK	NOT OK	List All Corrective Work Done **AND** report all Work Notifications written
	O	O Report	

Procedure Specific Tasks	OK	NOT OK	List All Corrective Work Done **AND** report all Work Notifications written.
As-Found	O	O	
As Left	O	O Report	

This PM performed by _____ **on** _____ .

Miscellaneous Starters/Contactors

Manufacturer		Model and Type	Various	Specific Equip. Data
Equipment Type	Electrical Switches			
Purpose(s)	1. Maintain component reliability, improve safety and increase production by eliminating unplanned downtime through properly planned and scheduled maintenance.			

Before proceeding with this work, the equipment needs to be safe. This means that all potential energy sources, such as pressure or power, need to be positively identified and locked out/tagged out according to your platform specific LO/TO procedures for this of equipment.

Take time to plan and think about your work.

Inform Operations (Platform Lead, PRC, or Field Foreman) before proceeding with the following maintenance activity.

For shutdown PM's, confirm that all potential energy sources have been identified, and LO/TO.

For all PM's, be aware of the other activities going on around you, and take necessary precautions to eliminate maintenance-induced failures.

Requirements	
Tools	⋎ Hand tools.
Consumables	
Spare Parts	Part Number

Step	Action	Notes, Diagrams and Key Notes
A *Preparation Visual*	1. Ensure system is safe to work on **OR** Ensure system is properly LO/TO at panel board. 2. Perform visual inspection (look for leaks, irregular sounds/noise, abnormalities and loose components). 3. Review drawings and identify relevant components. 4. Inform Production Operations of the tests to be performed.	
B-1 *Task*	1. Remove any unnecessary clutter around switch. 2. Inspect switch wall mounting bolts for tightness. 3. Actuate on/off switch to verify proper movement and contact connection.	*Note the following spare parts :* 3813 Fan 3810 Fin 1201 Air Compressor Air Filter 1202 Air Compressor Oil Filter

1203 Air Compressor Gauge
1804 Lube Oil Pump Filter

4. Inspect all indication lights to verify the bulbs are not burnt out and the lens covers are not broken and in place.
5. Verify all conduit connections are secure and properly terminated.
6. Inspect all power and control wiring terminations for tightness.
7. Inspect wiring for any signs of deterioration or excessive heating in the insulation.
8. Inspect contactor contacts for any sign of damage or pitting.

B-2

If ...	Then ...
	1. . .

Note :

B-3

B-4

If ...	Then ...

B-5	1.
C-1 *Return to Service*	1. When completed inform Production Operations that equipment is in service.
C-2 *Clean Up*	1. Clean up and remove all debris, parts and/or tools. 2. Dispose of discarded parts and/or contaminated parts as per platform HSE procedures.
D *Data*	Complete information where indicated on the Notification and Feedback form.	*Accurate reporting creates accurate equipment histories*

Installing V-Belts

1.0 PURPOSE: This procedure provides detailed instructions on how to install and maintain V-belt drives. The methodology described within this procedure is applicable to all V-belt driven installations.

2.0 SCOPE: This procedure supplements existing maintenance procedures.

3.0 EQUIPMENT:
  3.1 Mechanics Tools
  3.2 Approved replacement V- belts
  3.3 V-belt slip gages
  3.4 V-belt tension instrument
  3.5 Approved cleaning solvent
  3.6 Lint free cleaning rags
  3.7 Stroboscopic light tuned to operational speed of unit

4.0 SAFETY:
  4.1 Ensure coordination with operations
  4.2 Observe all site and area safety precautions at all times
  4.3 Ensure equipment has been isolated from all power sources and all LO/TO/TO procedures have been put in place
  4.4 In the case of fans, insure the inlet/outlet dampers are closed to prevent back flow from in service fans

5.0 PROCEDURE:
  5.1 Remove drive belt guard.
  5.2 Slowly roll the unit and inspect the belts and sheaves for the following:
       5.2.1 Side wall damage on sheaves.
       5.2.2 Belt surfaces for tears, cracking, and burn marks.
       5.2.3 Unusual stretching of belts.
       5.2.4 Worn side wall of v-grooves on sheaves.
       5.2.5 Ensure the belts are not bottomed out at the base of the sheave's v-grove.
       5.2.6 Inspect driving and driven keys and key ways for looseness and correct length of keys. NOTE: *This is an important inspection point. Incorrect sized keys cause unbalance in rotating equipment.*
  5.3 Detension v-belts by slacking off the drive motor hold down bolts. Then screw the tensioning bolts toward the driven component until the belts can be rolled out of the sheaves by hand. NOTE: *NEVER remove v-belts that are under tension by forcing them out of the sheave groves with a pry bar.*
  5.4 Using an approved cleaning solvent, thoroughly clean all groves in each sheave to remove dirt or oily film.
  5.5 Use the grove slip gage to inspect each grove in both the driver and driven sheaves for unusual wear patterns. Look for undercutting on groove side walls. If the side walls are barrel shaped, the sheaves should be replaced.
  5.6 Each grove must be inspected for nicks and scratches that could damage new belts. All defects found must be removed prior to installing the new belts
  5.7 Obtain a new set of replacement v-belts. Ensure the belts are correctly matched by carefully checking the manufacturer's code identification. NOTE: *NEVER mix v-belts with different manufactures' belts. Even if the belts match in size, there are different characteristics built in to the manufacturing process that could adversely affect their performance and longevity of operation.*

Figure C-17 V-Belt Installation

5.8   Verify that the alignment of the driven and driving components are aligned within 0.003 of an inch with each other. To determine alignment, place a straight edge diametrically across the outside face of the driven sheave and bring the straight edge to line up diametrically with the outside face of the driving sheave. If the components are correctly aligned to each other the straight edge will lie squarely on both sheave faces. If the straight edge does not lie square the driving component must be move until this is accomplished.

5.9   Inspect all of the new v-belts surfaces before installing the belts. Pay attention for cracks, fabric tears, oil impregnation spots, and verify that all of the belts are identical to each other.

5.10   Install the v-belts one at a time on to the sheave. Ensure the top, or outside edge, of the belt is sitting on the outer edge of the v-grove and the flat outer surface of the v-belt is flush with the outer circumference of the sheave groove.

5.11   Before tensioning the v-belts, ensure the motor hold down bolts are tight enough to prevent the motor from moving out of alignment, but will be free enough to permit the motor to slide slightly during the belt tensioning procedure.

5.12   Pull all the slack on the belt set to the bottom side of the belts. Gradually tension the belts by adjusting the tensioning bolts on the motor base. Periodically check the belts to ensure each belt is correctly tensioned. NOTE: *Slow rotation of the unit may be necessary midway through the tensioning process to ensure even and proper tension on all belts.*

5.13   If no belt tensioning tool is available follow this formula;

5.13.1   One inch of pitch (The distance between the centers of the driver and driven shafts' centerlines) = 1/64 of an inch deflection at the center of the span. Determine the deflection by placing a straight edge across the top of the belts, then push down on the top of the belts at the center of the span with a scale and measure the deflection.

EXAMPLE:
Pitch is 64 inches, the deflection is 1 inch
Deflection = Pitch x 1/64"
64" x 1/64" = 1" deflection

5.14   Verify drive and driven component alignment per step 5.8 .

5.15   Draw a straight tin line at 90 degrees to the rotation of the belts with a quick dry paint marker. Exercise extreme caution to avoid the rotating components. Start the unit and using a stroboscopic light tuned to the operational speed of the unit, determine the performance of the belts under tension by strobing the paint reference mark. If the belts are correctly tensioned and the components are correctly aligned there should be no appreciable movement of the painted reference mark.

5.16   Observe how the belts are running. Make any minor adjustment necessary. Lock down the motor's hold down bolts.

5.17   Replace the belt guard.

5.18   Clean entire work area and return all tools to the tool crib.

5.19   Inform operations that the maintenance is complete and close the work order.

5.20   Inspect the performance of the belts over a few days after installation. The belts will experience normal stretching. This stretching must be corrected by adjusting the tension bolts to the proper belt deflection. NOTE: *After any adjustment the alignment must by checked as described in step 5.8*

Index

misuse failures 78
motors 112-115, 159-62, 186-193
MTTR 85
muffler 147, 152

non-mandatory maintenance 62-63
non-pyramiding PM 63

Occupational Health and Safety
  Administration see OSHA
oil analysis 14, 82
on-off switch 158
operating task 36-40
operators 27, 34, 49-51, 103-111
OSHA 28, 47, 62-63, 65, 97, 102
outside contractors 52
overdue PMs 95-96
overload coil 157
overtime 96

part numbers 56
parts requirements 55-59
PdM techniques 14
performance 67, 85-102
performance frequencies 15
planning 14
plant geography 62
pneumatic cylinder 109, 143-148
pneumatic motor 110, 148-153
predictive maintenance 32-33
pressure relief valve 136, 140
preventive maintenance
  compliance 89-90
  costs 66-71, 90-91
  developing 11-15
  efficiency 92-93
  importance 1-10
  information 23-24
  inspections 112-161
  overdue 95-96

parts 55-59
performance 85-102
procedures 19-20
requirements 23-29
tasks 49-53, 55-59, 61-71, 73-75
training 35-36
types 31-33
sheets 100-101
proactive culture 2
proactive replacements 32, 34
preventive maintenance 19-21, 33-35
production performance reports 13
pump inlet 131, 137
pumps 135-136, 139, 219-200
pyramiding PM 63

quality 9, 18, 85

random failures 77-78
reactive culture 2
redundant equipment 97
refurbishings 32
regulatory agencies 28-29, 61
relay panel 157
reliability 33, 82
repairs 88-89
resources 52-53
return line 131, 137
root cause analysis 78
RTF 14
Run to Failure 14

scheduling 15, 61-71, 74, 100
sheaves 118
shut down 40-46
simplicity 18-22
skills 3, 49-53, 98-99
sliding maintenance 61, 90
sonics 14, 82
spares 57-59